The Wild Orchids of California

The Wild Orchids of California

Ronald A. Coleman

COMSTOCK PUBLISHING ASSOCIATES
a division of CORNELL UNIVERSITY PRESS
ITHACA AND LONDON

First published 1995 by Cornell University Press
First printing, Cornell Paperbacks, 2002

Printed in the United States of America
Color plates printed in Hong Kong

Library of Congress Cataloging-in-Publication Data
Coleman, Ronald A.
The wild orchids of California / Ronald A. Coleman.
p. cm.
Includes bibliographical references (p.) and index.
ISBN 0-8014-3012-7 (cloth : alk. paper) — ISBN 0-8014-8782-X (pbk. : alk. paper)
1. Orchids — California — Identification. 2. Orchids — California — Pictorial works. I. Title.
QK495.O64C64 1995
584′.15′09794 — dc20 94-38825

Cornell University Press strives to use environmentally responsible suppliers and materials to the fullest extent possible in the publishing of its books. Such materials include vegetable-based, low-VOC inks and acid-free papers that are recycled, totally chlorine-free, or partly composed of nonwood fibers. For further information, visit our website at www.cornellpress.cornell.edu.

Cloth printing 10 9 8 7 6 5 4 3 2 1
Paperback printing 10 9 8 7 6 5 4 3 2 1

To Jan, Joel, and Troy

Contents

Maps, Figures, and Tables

MAPS

FIGURES

TABLES

Preface

For many years I have enjoyed growing orchids in a small backyard greenhouse. Not long after acquiring my first commercially grown orchids, I learned that wild orchids grow in California and elsewhere across the land. That knowledge took me from my greenhouse to the mountains, valleys, and seashores of the Golden State. Since 1972, I have devoted vacations, weekends, and holidays to searching for, studying, and photographing California's wild orchids, both in the field and in various libraries and herbaria. This book, as well as a number of published articles, developed as a natural extension of that hobby.

There are two major treatments of the wild orchids of the United States, one by Correll (1950; reprinted 1978), the other by Luer (1975). These are supplemented by several regional orchid floras, none of which covers California. I decided to fill that gap by writing a book that would draw on my rapidly accumulating knowledge of the wild orchids of my home state.

The concept of the book evolved slowly, and expanded in scope several times. At first I intended simply to describe and illustrate the orchids that occur in California. In order to do that it was necessary to *know* which orchids grew in the state, and to learn where they were. This step initiated an extensive literature search during the initial phase of research, and none of the available references, I discovered, contains a complete list of California's orchids. The two major floras of California, those by Jepson (1951) and Munz (1968), mention only about 24 orchids each, but with slightly different lists. The orchid books by Luer and Correll cite several additional species and varieties, and one species was reported only by Sorrie (1978). When he revised the genus *Piperia*, Ackerman (1977) added several more species to the list of those in California.

The list thus continued to grow, and four new species were described only after preparation of the book had begun.

As the text for this volume was being completed, a new California flora, *The Jepson Manual: Higher Plants of California* (Hickman, ed., 1993), was published. Completed under the auspices of the Jepson Herbarium at the University of California, Berkeley, the volume contains a then current listing of the wild orchids in California, although the nomenclature differs slightly from that used here. The orchid treatment in *The Jepson Manual,* the work of Dieter H. Wilken and William F. Jennings, will be noted in the species accounts whenever their nomenclature differs from that used in *The Wild Orchids of California.*

I found, too, that different authors employ different nomenclature, or names of species. *The Wild Orchids of California* follows the nomenclature of Luer for all genera except *Piperia.* Because Morgan and Ackerman (1990) and Morgan and Glicenstein (1993) have done the most recent work on that genus, this treatment follows their lead. The literature is also replete with a scattering of varietal names that various authors have used to describe distinctive populations of a species. In the chapters devoted to the separate genera, the varietal names in the literature are discussed and, where appropriate, my reasons for using them or not using them are given. Nomenclature differences with Hickman, ed. (1993) are mentioned in the species texts.

While attempting to select appropriate illustrations for the species, I realized that my slide collection, though large, did not adequately record all of the wild-orchid species in California. Some slides, such as those of *Platanthera stricta* and *P. hyperborea,* had been taken in other states, and could not be used to document the presence of the plants in California. In other cases, my slides were not quite good enough for publication. A multi-year project ensued to find and photograph every species and variety of wild orchid in the state, and to document color and size variation as well. Old sites were revisited, and new ones were found. A call from a native-plant lover or forest ranger would trigger a 1200-kilometer weekend trip to photograph an orchid in bloom. Finding a suitable plant sometimes took multiple searches over several weeks. On other occasions we arrived a little too soon or too late for the blooming season, and had to adjust the timing of our visits the following year. Getting the photographs was the hardest and most time-consuming task of this project, but in a way also the most rewarding, because the beauty of the flowers is captured forever. All of the reproductions in this book are from living plants photographed in their natural habitats.

Early in the research phase of this work, I decided to document the statewide distribution of the orchids. Herbarium collections proved indispensable for this task. In addition to documenting distribution, the collection records were also the best source for blooming season and elevational data. Initially, I visited individual herbaria, either making a special trip during the off-season, or stopping by as part of an orchid-hunting field trip. After a few years it was apparent the book would never be completed if I were to visit all of the pertinent herbaria personally. My herbarium studies were accelerated and greatly aided through an association with the Rancho Santa Ana Botanic Garden. Thomas Elias, the director, and Steven Boyd, the herbarium manager, arranged for me to have weekend access to their facility, and later requested loans of orchid collections for me to examine, from herbaria throughout the state. These steps saved untold numbers of travel hours and greatly shortened the time needed for herbarium studies. Appendix 1 contains a complete list of the herbaria that were consulted, either by visits or by loans of their collections to the Rancho Santa Ana Botanic Garden. The main body of Appendix 1 is a tabulation summary of distribution data. For each orchid species, each county in which it occurs is identified by an entry listing an herbarium that contains a representative collection from that county. To keep Appendix 1 from becoming too large, only one entry is made for each species per county, but in most cases, multiple collections, from several herbaria, are available from each county.

For those unfamiliar with an herbarium, a few words of explanation may be useful. As they have for centuries, herbaria store pressed and dried plants for scientific study. The plants are collected by researchers (or, in the past, even capable amateurs), mounted on "sheets," and sent to the herbarium for preparation and retention. With each plant the collector cites pertinent information such as date, location, elevation, and habitat. The pressed specimens provide both a historical reference to the flora of an area and a source of data for researchers. With access to existing scientific collections it was not necessary for me to visit personally every county multiple times during the blooming season to determine what orchids grew there. Whenever the text refers to a collection, it is to one of these dried and pressed specimens stored in an herbarium, and not to collections in the horticultural sense of removing plants from the wild to be grown in gardens. I neither advocate nor practice the removal of any of our native flora.

The last extension of the scope of the original project, settled on in 1990, was to document the available data on the pollination mechanisms of the wild orchids in California. This decision initiated another flurry of library study. Various researchers have studied the pollination biology of our orchids, but for the most part the information is widely scattered. Some of the earliest but still valuable research on orchid pollination was done by Darwin (1877). Although he did not examine any of the California species, he did study Old World representatives of some of the genera represented in California, and subsequent researchers have shown that his conclusions apply to our plants also. Data of this sort are referenced in the genus treatments; data specific to particular species are discussed in the species treatments. Catling and Catling (1991) generated a table summarizing references on pollination for all North American native orchids. They found, as I did, that the studies are incomplete, and that the pollination biology for some of our wild orchids has not yet been documented.

Many wildflower books and orchid books devote sections to cultivation, but I have deliberately avoided any reference to the cultivation of these plants. In general, the wild orchids in California are not available commercially, and there are no firms offering seed-grown plants for sale. The literature contains many papers on the work that universities, colleges, and botanical gardens have conducted in an attempt to germinate and grow the North American temperate-zone wild orchids. Although results have been achieved with some species (see Olive and Arditti, 1984), the limited technical progress has yet to be transferred to major commercial operations. Moreover, the cultivation requirements of most of these orchids have not been defined. *Epipactis helleborine*, a notable exception, germinates readily from seed scattered in gardens. But given that no seed-grown plants are available, the only source for these or other orchids is to remove them from the wild, which violates state and local ordinances. Since growing techniques are still being developed, the digging up of wild plants almost certainly dooms them to an early death. Even after all these years, I still feel an emptiness on discovering a hole in the ground where an orchid once grew. The diggers invariably seek the most beautiful plants, and well-meaning flower lovers have sent uncountable orchids to an early grave. For those seeking to grow these plants, the best advice is, "Don't try!" Take a walk in the forest or the fields to see them in nature, and take as many photos as you like, but leave the plants for the next generation of orchid-lovers to enjoy. Happy orchid hunting.

A work of this type would not be complete without acknowledging the many people who helped with one or another aspect of the task. Many thanks to Paul Gripp, who recognized my early interest in the wild orchids, and continues to call with encouragement and new information. Thanks also to Leon Glicenstein, J. "Hawkeye" Rondeau, and Edith Rondeau, who spent many weekends with me hiking the trails and roaming through the woods and fields searching for elusive varieties. They also sent regular updates on their own treks in search of wild orchids. Randall Morgan provided invaluable help with *Piperia,* and also spent many hours hiking and studying herbarium collections with me. Paul, Leon, Hawkeye, and Randall also generously reviewed all or portions of the text. W. Juergen Schrenk, formerly of Abbott Laboratories, Charles Sheviak of the New York State Museum, and John Atwood of the Orchid Identification Center at the Marie Selby Botanic Garden all answered questions and shared information. Thomas Elias and Steven Boyd of the Rancho Santa Ana Botanic Garden provided enormous assistance by arranging for me to use their magnificent research institution, and by borrowing collections for me to study. James D. Ackerman of the University of Puerto Rico reviewed the text several times, offering welcome suggestions on organization, style, and content, and making much needed technical corrections. I am also grateful to members of the American Orchid Society and the California Native Plant Society too numerous to mention, who told me of orchids in bloom, or of additional locations where they might be found.

Occasional help came from unsuspected quarters. Many other orchid-lovers apparently do as I do and look for wild orchids wherever they happen to be, and several orchid-hunters, some from other states and even foreign countries, took the time to write to me about discoveries they had made while orchid-hunting in California.

My greatest thanks go to those who provided the most support: my wife, Jan, and our boys, Joel and Troy. All three made many orchid-hunting trips with me during the last 10 years, and over the last 20 years Jan made most of them. The extra sets of eyes often detected a hidden plant I had already passed. We searched in the rain, in the cold, and in the heat, often in difficult terrain, and we put up with some primitive camping conditions, all in the name of completing the slide collection. The trips the boys did make—they were still of school age—were often driving and hiking marathons, when most of the time was spent in the car or on the trail, rather than on the things boys prefer. On weekends when I was not in the field, I could be found in libraries or herbaria studying literature or specimens, and during evenings I would be making notes on the research or creating drafts of the text. With a maturity belying their years, the boys accepted my obsession, and with an understanding beyond comprehension, Jan bore the extra parenting load while I worked on the book. To them, then, this book is dedicated, with the same love and devotion they showed for me.

RONALD A. COLEMAN

Tucson, Arizona

The Wild Orchids of California

1. Introduction

From the ridge in the western reaches of the Santa Monica Mountains the trail wound sometimes gently, sometimes steeply, down the burned-out canyon. On the far side it showed faintly as a light line on the ash-darkened landscape. Still, there was green all around. This first spring after the fire, life was returning to the scorched earth. A profusion of wildflowers danced in the breeze, completely filling newly opened areas, even crowding under the remnants of burned chaparral. The chaparral, too, was recovering: new shoots were emerging on the surviving crowns of the torched shrubs. The tiny stream at the base of the canyon still bubbled and gurgled in tribute to the winter's rains, and crossing it without getting wet took a mighty leap. The stream would still be flowing later in the summer in all but the driest years, but by then it could be crossed without breaking stride. In the canyon bottom, recovery was even more advanced than on the slopes, but dead oak trees were mute testimony to the intense heat of the fire. On the western side of the canyon the trail turned uphill, following the stream. It climbed gradually, twisting to traverse small seasonal flows trickling down to join the main stream. About 20 meters past one of these drainages, almost in the shadow of a burned but still-alive yucca plant, was the object of the search. Inconspicuous among the grasses, and almost the same shade of green, were two blooming plants of the tiny orchid *Piperia cooperi.*

Although *P. cooperi* is fairly rare, other orchids in Southern California are not. In fact, *P. cooperi* is just one of two orchids that share the common name chaparral orchid because they grow in the dry, chaparral-covered hills so widespread in the southern part of the state. The other one, *P. elongata,* is also a *Piperia.* Most people have no trouble visualizing an orchid, though they would probably picture the commercially grown flowers once reserved for special occasions but now becoming more frequently used as cut flowers. Some might remember that moccasin flowers or lady's slippers are members of the orchid family, but only the rare lover of native wildflowers would know that *P. cooperi* is an orchid. Part of the lure of the orchids has always been their ability to amaze us with their beauty and their intricate, varied, flower structure, characteristics not only of *P. cooperi,* but of all the wild orchids in California.

Van der Pijl and Dodson (1966) estimated that orchids make up nearly 7% of all species of flowering plants. Estimates of the number of orchid species range from 15,000 to 35,000, with more being identified every year. The vast majority grow in tropical or subtropical climates, but with the exception of Antarctica, orchids are found on every continent, even growing within the Arctic Circle. Luer (1972, 1975) described over 200 native orchids in the United States. Many of these are subtropical species that do not grow north of Florida, but approximately 110 of the native orchids within the United States are temperate and occur *only* north of Florida. In California we have 32 species in 11 genera.

What Is an Orchid?

Before studying the details of California's native orchids, we should perhaps consider just what botanical characters cause us to say, "This is an orchid." Many orchid books, especially those by van der Pijl and Dodson (1966) and by Dressler (1981), give excellent definitions of an orchid. Interested readers are referred to either of those works for a more detailed discussion of orchid characteristics than is presented here. Dunsterville and Dunsterville (1986, p. 604) say, "It also has not been easy to be sure that we have outlined [just] how to distinguish an orchid, because the composition of the orchid is so involved that, as will be described later, there is no one (or even two) features that you can pin down as truly diagnostic." In the view of many botanists, in fact, orchids are the most highly evolved of all plants. As members of the subclass Monocotyledoneae (the monocots), they are related to, among others, the lilies, irises, bromeliads, palms, sedges, rushes, and grasses.

What follows is a summary of the features that can be used to identify a flower as an orchid (definitions of the botanical terms employed here can be found in the Glossary):

First, in most flowering plants the stamens and pistil are separated, but in the orchids they are fused together into a single complex structure called the *column,* which is usually presented opposite the *lip,* or lower petal. Orchids produce lumped pollen, called *pollinia* (plural of *pollinium*), which can be anything from loosely bound grains to hard lumps. The *anther cap,* at the tip of the column, covers and protects the pollinia. A sticky element called the *viscidium* is usually connected to the pollinium, sometimes with a connecting structure called the *stipe,* or an extension of the pollinia called the *caudicle.* The combination of pollinium, viscidium, and stipe or caudicle is called a *pollinarium.* Visiting insects, typically bees or flies, come in contact with the viscidium in the act of foraging, and it sticks to them immediately, withdrawing the pollinia as the insect retreats. The *stigma,* atop the pistil, is coated with a sticky substance to secure the pollinia deposited by the next visitor. The stigma, in orchids just a slight depression in the column, is formed by the union of the three *stigmatic lobes* found in monocotyledons. In the orchids a portion of the third lobe is usually modified into a *rostellum,* the flaplike structure situated between the pollinia and the stigma. Part of the function of the rostellum is to serve as a barrier between the pollinia and the stigma to prevent self-pollination. Because of the relative positions of the column parts, the pollinator picks up the pollinia as it exits the orchid flower, then deposits them on the stigma of the next flower visited, effecting pollination. And once again, new pollinia are picked up as it leaves.

Second, orchid seeds are produced in great numbers. Some species produce millions in a single mature ovary, or *capsule.* The seeds are minute, dustlike, and unlike most other plant seeds they carry little or no food supply for the emerging plant.

Third, the *outer whorl* of the *floral envelope* of an orchid consists of three *sepals.* The uppermost sepal is called the *dorsal sepal,* the lower two the *lateral sepals.* All three are usually the same color and shape as the upper two petals, though they may differ from the petals in both size and shape. Sometimes, as in *Cypripedium,* the two lateral sepals are joined into a single structure called the *synsepal.*

Fourth, the *inner whorl* of the floral envelope comprises the *petals,* and two of them, the upper two, often resemble the sepals. The third, or lower, petal has been transformed into an often elaborate element called the *labellum,* or *lip.* The lip is usually involved in pollination, as a landing platform or attractant device for the pollinator, and as such is often of a different color than the sepals and petals, and is usually much larger and shaped quite differently. It can have ridges, deep clefts, side lobes that resemble teeth, or an array of hairs, or it can be shaped like a pouch or slipper. The lip often incorporates a downward- and backward-directed tube, or *nectary,* called the *spur.* The spur sometimes, but not always, contains nectar. Nectar production is not limited to the spur; some species also secrete it from other parts of the lip, and some produce no nectar at all. Many orchids produce a faint to strong scent, as an additional attractant to pollinators.

Fifth, as the flower bud of an orchid matures, the stem of the flower, or *pedicel,* gradually rotates so that the lip, which starts out on top, is presented at the bottom when the flower opens. This process is called *resupination.* Most orchid flowers rotate through 180°, but some rotate a full 360° and others do not rotate at all. The effects of resupination are often easily observed by studying the twisted pedicel and, in some cases, ovary.

Sixth, the orchid flower has an *inferior ovary*—that is, the sepals and petals are attached to the apex of the ovary, rather than at its base.

Finally, the overall arrangement of the orchid flower results in *bilateral* (not radial) *symmetry.* Thus, a line from the tip of the dorsal sepal to the center of the lip bisects the column, and the two sides of the flower thus bisected are mirror images of each other.

These characteristics, or *characters,* can be used to identify a flower as an orchid (Figure 1); not all orchids, however, have all of the features mentioned. For example, some do not have a rostellum, and the column structure in *Cypripedium* differs from that described above. Conversely, some of the features mentioned above are shared with other families. For

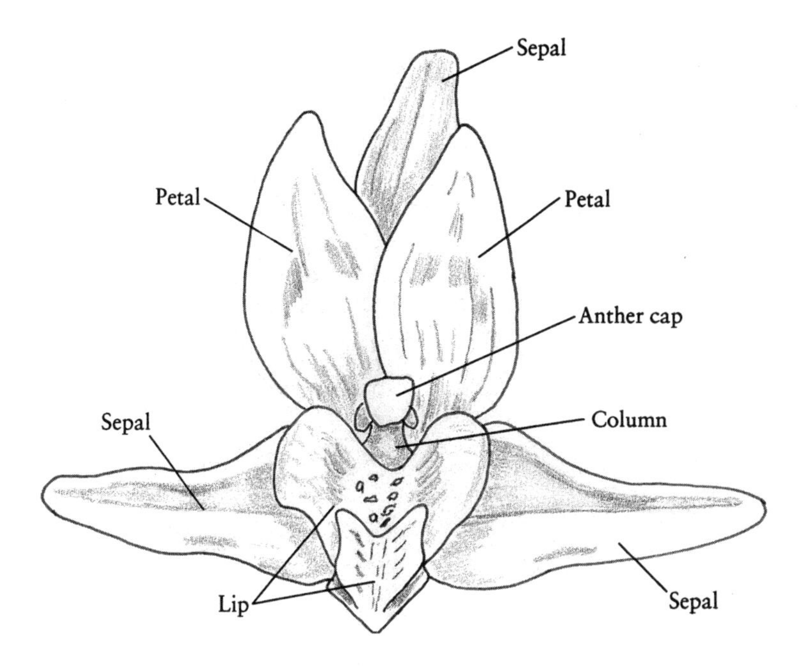

Figure 1. Parts of the orchid flower

example, members of the Asclepiadaceae family (milkweeds) have a structure similar to the column of orchids; members of the Liliaceae have sepals and petals of similar size and shape; and members of the Iridaceae have an inferior ovary, as do orchids.

The plant structure of orchids defies characterization to a single or even a few types. All of our species in California are nonbranching and have a sympodial growth habit—the next-year's growth emerges from a lateral bud of the current growth—but are otherwise varied. The leaves of all California orchids are parallel-veined (in some orchids elsewhere, they are not) and either plicate or conduplicate. Leaves may be basal or scattered along the stem. Some species have single leaves, others have dual leaves, still others have multiple leaves, and several have no leaves. With experience it is possible to identify the vegetative stage of a plant as an orchid, but in those cases where there are multiple species in the same genus, particularly in *Piperia, Platanthera,* and *Spiranthes,* species identification based on leaf character alone is uncertain.

Part of the amazement and beauty of the orchids lies in the myriad methods they use in presenting their structures. Van der Pijl and Dodson (1966) summarized orchid variation by saying, "There is probably no other plant family which demonstrates such remarkable variation in the form and functions of the basic components of the flower. They are infinitely varied and often in a progressive manner." Look for these variations as you study the illustrations included with the species descriptions, or while you examine the plants in the wild.

Blooming Seasons and Distributions of the California Orchids

Students of native orchids usually have three basic questions about the family in California: Which species occur in the Golden State? Where do they grow? And when do they bloom? Here I shall attempt to answer those questions by presenting an overview of names, blooming seasons, elevational distribution, and geographic distribution (Table 1).

The blooming seasons are presented graphically in Figure 2, which is intended as an easy reference for those wanting to search for California's wild orchids. A quick glance shows the species expected to be in bloom during a planned visit. The combined blooming season for the state's orchids stretches over nine months, from February to October, allowing us to enjoy their beauty for much of the year. For those interested in seeing the maximum numbers of wild orchids, rather than specific plants, May, June, and July are the peak months for orchid bloom. The blooming seasons for given species (detailed in the species texts) are defined by the earliest and latest dates I have established, whether from my own field research or from the dates when herbarium collections were made. The actual blooming seasons may extend somewhat beyond those shown, because it is unlikely that my visits or the herbarium collections took place on the exact first or last day of bloom. It is also just as likely that, owing to a late spring or early fall in a particular year, the season could start later or end sooner than shown in Figure 2. The start of the blooming season for a given species can thus shift as much as three weeks from year to year. Therefore, if you are planning to search for a particular orchid, schedule your search near the middle of the blooming season.

Table 1. Blooming seasons and distribution, by elevation and county, of California's wild orchids

Species	Blooming season		Elevation range		Range limits, by county name	
	Begins	Ends	Low (m)	High (m)	Southern limit	Northern limit
Calypso bulbosa	2 Mar	14 Jul	0	1798	Santa Cruz	Del Norte
Cephalanthera austiniae	3 Mar	3 Aug	0	2195	San Diego	Del Norte
Corallorhiza maculata	22 Feb	11 Aug	15	2743	San Diego	Del Norte
Corallorhiza mertensiana	6 May	7 Aug	30	2195	Sonoma	Del Norte
Corallorhiza striata	19 Feb	30 Jul	76	2210	Fresno	Siskiyou
Corallorhiza trifida	25 Jun	4 Jul	1372	1676	Plumas	Plumas
Cypripedium californicum	11 Apr	28 Jul	61	2134	Sonoma	Del Norte
Cypripedium fasciculatum	12 Mar	31 Jul	171	1981	Santa Cruz	Del Norte
Cypripedium montanum	1 Apr	9 Jul	183	2134	Madera	Del Norte
Epipactis gigantea	3 Mar	1 Oct	0	2591	San Diego	Del Norte
Epipactis helleborine	14 Apr	6 Sep	30	1219	Monterey	Sonoma
Goodyera oblongifolia	6 May	19 Oct	0	2134	Tulare	Del Norte
Listera caurina	25 Apr	18 Jul	30	1966	Humboldt	Del Norte
Listera convallarioides	24 May	23 Aug	762	2896	Riverside	Del Norte
Listera cordata	21 Mar	22 Jun	46	610	Mendocino	Del Norte
Malaxis monophyllos	12 Jul	22 Aug	2225	2652	Riverside	San Bernardino
Piperia candida	27 May	9 Sep	46	1195	Santa Cruz	Del Norte
Piperia colemanii	29 Jun	9 Aug	1219	2286	Tulare	Siskiyou
Piperia cooperi	27 Mar	6 Jun	0	914	San Diego	Ventura
Piperia elegans	17 May	27 Sep	0	533	Santa Barbara	Del Norte
Piperia elongata	8 May	16 Sep	0	2073	San Diego	Del Norte
Piperia leptopetala	16 May	31 Jul	381	2134	San Diego	Siskiyou
Piperia michaelii	1 May	18 Aug	15	914	Los Angeles	Humboldt
Piperia transversa	25 May	23 Aug	24	2073	San Diego	Del Norte
Piperia unalascensis	2 May	15 Aug	122	2590	San Diego	Del Norte
Piperia yadonii	23 Jun	12 Aug	30	61	Monterey	Monterey
Platanthera dilatata	9 May	11 Sep	3	3353	San Diego	Del Norte
Platanthera hyperborea	3 Jun	21 Aug	1280	3139	Inyo	Siskiyou
Platanthera sparsiflora	15 May	1 Sep	116	3353	San Diego	Del Norte
Platanthera stricta	10 May	15 Aug	1067	2286	Humboldt	Del Norte
Spiranthes porrifolia	27 May	8 Sep	30	2499	San Diego	Del Norte
Spiranthes romanzoffiana	18 Jun	30 Sep	0	3261	Riverside	Del Norte

California's orchids grow in extremely diverse habitats from sea level to at least 3350 meters (Figure 3). Spring comes later the higher one gets in the mountains, and orchid hunters should allow for differences in blooming due to elevation. Early in the season, search the lower elevations, and near the end of the season, look higher.

The numbers of orchid species occurring in the various counties of

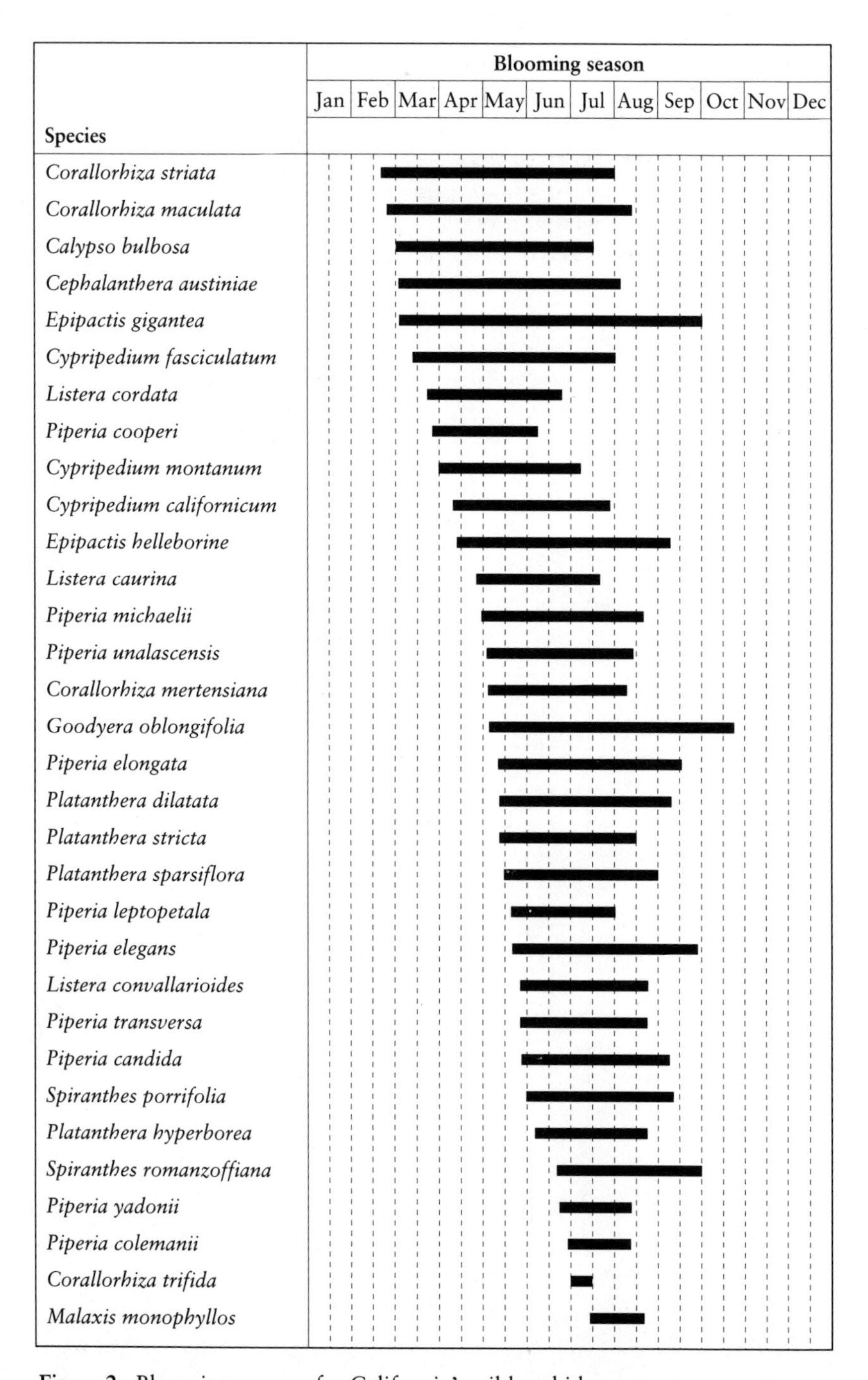

Figure 2. Blooming seasons for California's wild orchids

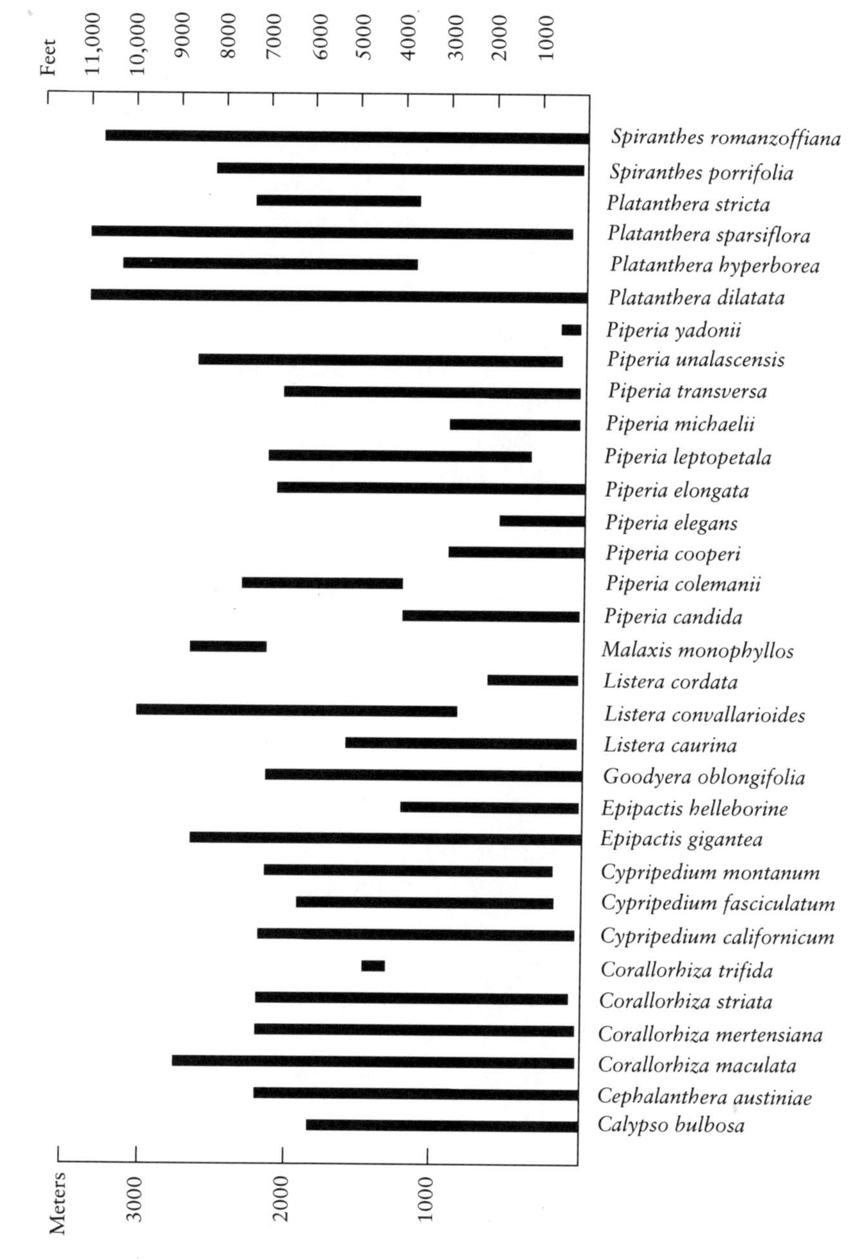

Figure 3. Elevational distribution of California's wild orchids

California varies considerably, depending on such factors as rainfall and topography (Map 1 and Table 2). Some orchids are found in almost all of the state's 58 counties (Map 2 names the counties and locates the Channel Islands). The highest concentration of orchid species is in the wet northwestern counties of Siskiyou, Del Norte, Trinity, and Humboldt, and the fewest orchids occur in the dry, flat Central Valley. The next-best regions for orchid diversity are the mountainous counties at the northern end of the Sierra Nevada range and the coastal counties as far south as Monterey. The concentration of orchid species thins out in Southern California, except for the surprisingly high numbers in San Bernardino and Riverside Counties, which are due to the Sierran climate of the San Bernardino and San Jacinto Mountains. San Diego County, too, because of its mountains, supports a respectable number of orchid species. Some counties—Alpine, Amador, and Yuba, for example—seem to be underrepresented in species. The counties adjacent to these have several more species, although the array of habitats is virtually identical. This anomaly should be corrected (or explained) by additional field studies in those counties. Conversely, examination of the distribution maps for the various species will disclose apparent gaps in our knowledge of the exact range of these species within California. The documentation for additional range may be available somewhere, but it does not appear in the herbarium records I studied. Or, the records may simply be incomplete, and additional field searches may yet expand the known range for some of our species, or even add new species to our flora. Appendix 2 offers my speculation about possible additional locations for many of our orchids, as well as two potential new orchids for California. The reader is challenged to search the counties listed in the hope of gathering new data.

Two orchids, *Corallorhiza trifida* and *Piperia yadonii,* have a precarious hold in California, each occurring in just one county (Table 3). The former is found only in Plumas County (but occurs also in other states), and the latter is endemic to Monterey County (that is, it occurs nowhere else in the world). *Malaxis monophyllos var. brachypoda* is historically known from two counties, but may be extant in only one. At the other end of the scale, *Epipactis gigantea* occurs in 46 counties and may eventually be found in several more.

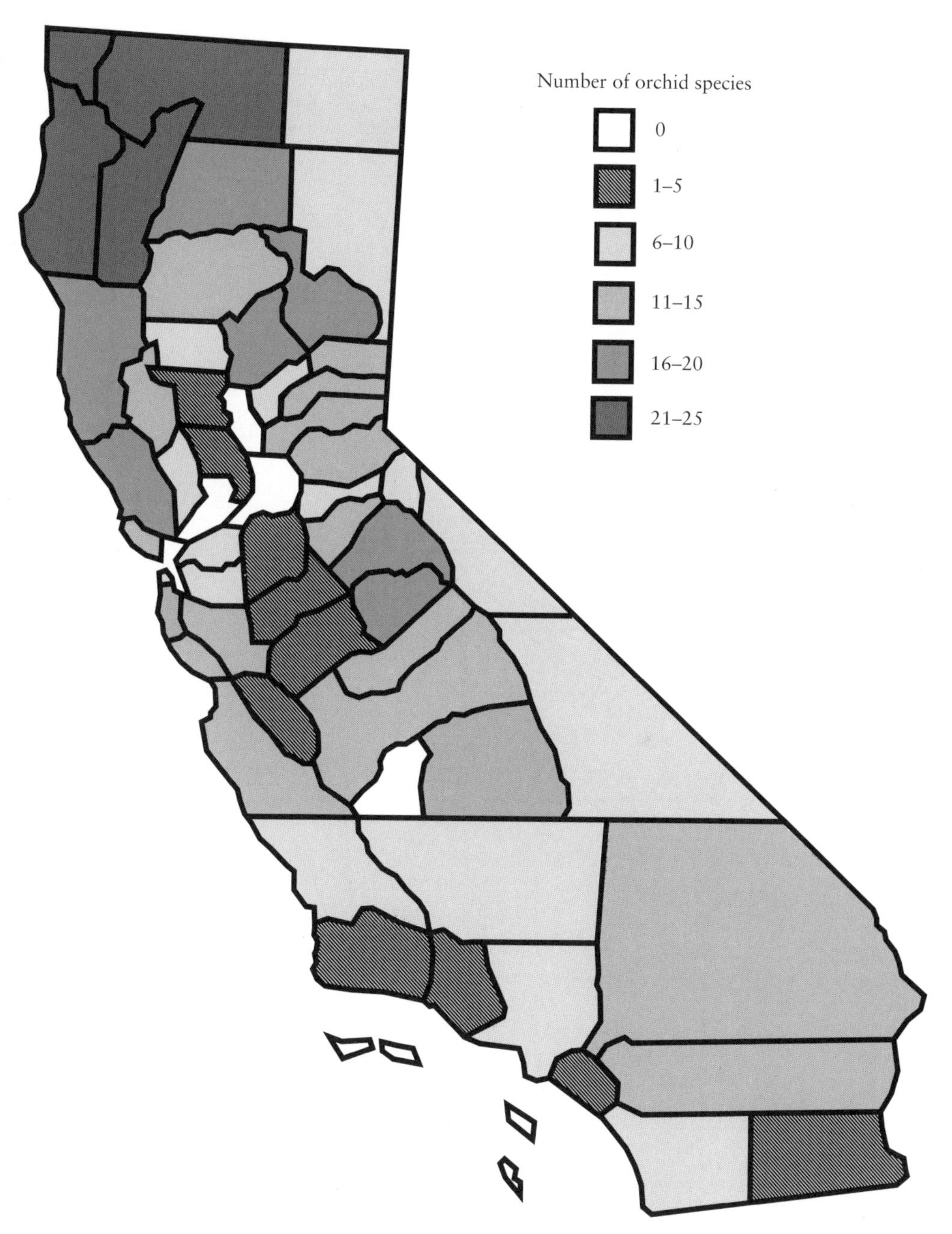

Map 1. Distribution of California's wild orchids, by number of species per county

Map 2. The counties of California

Table 2. Distribution of California's wild orchids, by number of species per county (as well as Channel Islands)

County	No. of species	County	No. of species
Alameda	6	Orange	3
Alpine	7	Placer	12
Amador	9	Plumas	20
Butte	17	Riverside	11
Calaveras	13	Sacramento	0
Channel Islands	3	San Benito	5
Colusa	4	San Bernardino	14
Contra Costa	8	San Diego	10
Del Norte	23	San Francisco	6
El Dorado	14	San Joaquin	1
Fresno	14	San Luis Obispo	8
Glenn	9	San Mateo	16
Humboldt	25	Santa Barbara	4
Imperial	1	Santa Clara	13
Inyo	7	Santa Cruz	15
Kern	6	Shasta	17
Kings	0	Sierra	13
Lake	14	Siskiyou	24
Lassen	8	Solano	0
Los Angeles	8	Sonoma	18
Madera	14	Stanislaus	2
Marin	15	Sutter	0
Mariposa	16	Tehama	15
Mendocino	20	Trinity	21
Merced	1	Tulare	16
Modoc	10	Tuolumne	16
Mono	9	Ventura	5
Monterey	14	Yolo	1
Napa	6	Yuba	9
Nevada	15		

Several of our wild orchids are California endemics—they occur naturally nowhere else in the world—and are restricted to certain geographical or climatic regions. Other species that occur here also grow elsewhere, but either were originally discovered here or were described from material collected in the state. For all those species that are endemic to the state, and for those described from collections within California, the original collector and collection locale are given.

Part of the thrill of my research in preparation for this book was dis-

Table 3. Distribution of California's wild orchids, by number of counties per species

Species	No. of counties
Calypso bulbosa	11
Cephalanthera austiniae	28
Corallorhiza maculata	41
Corallorhiza mertensiana	9
Corallorhiza striata	25
Corallorhiza trifida	1
Cypripedium californicum	9
Cypripedium fasciculatum	13
Cypripedium montanum	15
Epipactis gigantea	47[1]
Epipactis helleborine	10
Goodyera oblongifolia	26
Listera caurina	4
Listera convallarioides	30
Listera cordata	3
Malaxis monophyllos	2
Piperia candida	8
Piperia colemanii	11
Piperia cooperi	7[1]
Piperia elegans	17
Piperia elongata	35[1]
Piperia leptopetala	16
Piperia michaelii	20
Piperia transversa	32
Piperia unalascensis	36
Piperia yadonii	1
Platanthera dilatata	40
Platanthera hyperborea	7
Platanthera sparsiflora	31
Platanthera stricta	5
Spiranthes porrifolia	35
Spiranthes romanzoffiana	36

[1]Total includes occurrence on Channel Islands.

covering the population dynamics of California's native orchids. *Corallorhiza trifida* var. *verna* was not reported in California until 1977. *Malaxis monophyllos* var. *brachypoda* was rediscovered in 1989 after not having been observed for 42 years. *Piperia candida, P. yadonii, P. colemanii,* and *P. elegans* ssp. *decurtata* were only recently described. It is intriguing to think that other species, varieties, or hybrids are still waiting to be discovered, and that the number of known species in California is likely to increase once more.

But there is also a sad finding in the research: the range of several of our orchids is shrinking. The locations indicated on many of the herbarium collections no longer support orchids because they now support freeways, malls, dams, or homes, or have been clear-cut for timber. For example, the Los Angeles County distribution of some members of the genus *Piperia* has been greatly reduced, owing to destruction of habitat for development. Loss of range is also occurring for orchids in other areas of the state. *Malaxis monophyllos* var. *brachypoda* may have been extirpated from Riverside County; *Cypripedium californicum* has probably disappeared from Marin County; and repeated recent searches have failed to find *C. montanum* or *C. fasciculatum* in their historical range in the Santa Cruz Mountains. Perhaps these plants survive in some habitat niche throughout their original range, and will one day reappear, but the shrinking of their perceived ranges emphasizes the need for continued efforts in support of habitat protection before it is too late for them, or for others of our orchids.

The Biogeography and Climate of California

A thorough study of the wild orchids of California requires that we consider briefly the conditions in which they grow. Because of its size and its mixed topography, the growing conditions in California vary greatly. Within its boundaries are some of the most diverse growing conditions in the United States. As amazing as it may seem, orchids of one species or another grow in the most extreme environments California has to offer. They grow along desert springs, in the rain forests of the north, in lush Sierran forest, in Alpine meadows, and in the searing chaparral. Only four of the state's 58 counties—three of them in the Sacramento area—fail to sustain wild orchids (see Table 2).

The flora of California, too, is remarkably diverse. There are over 5000 native species of vascular plants in 875 genera in the state. Raven (1988) says "This region clearly contains the most remarkable assemblage of native plant species in all of temperate and northern North America." Raven estimates that 19 of California's genera and 1525 of its species are endemic. For all this richness in total plant population, however, the orchid flora in California is not as large as that in other parts of the country, but its members are beautiful and varied nonetheless.

California's southern border, with Mexico, is at about 32° north latitude, and is roughly 1320 kilometers from its northern border, with Oregon, which is at about 40° north latitude. The state is much narrower than it is long, just 552 kilometers east to west at its widest point. The Pacific Ocean forms the entire western boundary, Nevada and Arizona the entire eastern boundary. Elevations in California range from a low of 86 meters below sea level in Death Valley to 4418 meters above sea level at the top of Mt. Whitney. Total area within the state is slightly over 400,000 square kilometers, greater than all of Japan. With so great an expanse, no single description can adequately portray the varieties of its climate. Death Valley is the hottest and perhaps the driest spot in the United States. It averages only 5 cm of rain per year, and summer temperatures reach a sizzling 57°C. At the other extreme of precipitation, the mountains in the northwest receive tremendous amounts of rain, up to 235 cm per year. Though snow is rare in coastal California, it is common at higher elevations in winter, and several mountains support permanent glaciers.

Much of coastal California enjoys what is usually referred to as a Mediterranean climate, typified by damp, mild winters and dry, cool summers. Away from the coast, summers are usually warm to hot, except at the higher elevations in the mountains. Rainfall occurs mostly between November and April, although rain and thunderstorms may happen at any time of year in the mountains, and hikers should plan on them. Multi-year droughts occur every 10 to 15 years. Plant life in California is well adapted to this wet-dry cycle, and many of the plants, including some orchids, go dormant during the driest part of the year. Generalizations about climate, however, in a region as large as California are apt to be misleading. Both rainfall and temperature can vary greatly within a short distance, owing to elevation change or relative proximity to the coast. Those interested in a more extensive discussion of the climate in California, and its relationship to vegetation, should consult Major (1988).

A comprehensive treatment of the vegetation communities of California is that by Küchler (1988), who identifies 54 different types, many with subtypes. He identifies the plant communities by dominant growth forms, such as trees, shrubs, grasses, etc., and the dominant genera and species present. This seemingly large number of distinct plant communities allows recognition of the effects of both elevation and latitude. For example, Küchler distinguishes a redwood forest community, both northern and southern Jeffery pine forests, and four types of yellow pine forests in a total of 28 major forest communities. The influence of the Pacific Ocean on the vegetation in California is recognized by multiple coastal plant communities on the mainland, from salt marsh and coastal scrub to coastal forest; two of the orchids endemic to California grow *only* within these plant communities influenced by the ocean. Other major vegetation types recognized with multiple plant communities are the desert, chaparral, savanna, marsh, creosote bush, and scrub woodland. At least two of the plant communities—the hot sandy desert and the cold alpine desert, which are defined as being largely devoid of vegetation—exclude orchids altogether. In most of the other plant communities, however, there is a chance of finding at least one orchid species, either because it is a natural part of that community or because it persists in a particular ecological niche within that community. For example, Küchler defines two plant communities on the Channel Islands. Eight major islands make up that chain, and orchids are found on several of them. The *Piperia* species that grow there are natural members of one or the other of the two communities, or both. The stream orchid, *Epipactis gigantea,* also grows there, in wet areas around seeps or springs. It may not be a natural member of the community, but exists within the confines of the community because its particular conditions are met at the seeps. This happy juncture of general and special circumstances is present in most regions of the state, and with only a few exceptions, the assiduous orchid hunter can hope to find orchids in almost all of the plant communities in California.

Principal Geographic Features of California

Excellent descriptions of the geography of California are available in many books, and interested readers should consult M. Hill (1984), R. B. Hill (1986), and Barbour and Major (1988) for detailed discussions.

The brief descriptions offered here are intended to provide a general reference for some of the locations mentioned in the chapters devoted to the orchid genera.

Most authorities divide the mountains of California into eight major regions, as shown on Map 3, which is modeled after maps in M. Hill (1984), R. B. Hill (1986), and Barbour and Major (1988). The eight mountainous regions are called the Klamath, Cascade, Sierra, North Coast, South Coast, Transverse, Basin, and Peninsular Regions. In the aggregate, these regions contain over 100 mountain ranges. The number of independent ranges in each region varies greatly, from just one in the Cascade Region to 50 in the Basin Region.

The Klamath Region extends from Oregon down into the northwestern corner of California. This is the wettest area of the state, with portions properly called northern rain forest. Some of the major ranges in this region are the Trinity Alps, the Marble Mountains, the North Yolla Bolly Mountains, and the Siskiyou Mountains. Of our eight regions the Klamath is the richest in orchid species. It is also a wild and spectacularly beautiful region, with five major (often fast-flowing) rivers and dense forests. The timber industry is a major employer in the area, and each year much prime orchid habitat is lost to logging. The southeastern end of the Klamath Region encompasses the jagged spires of Castle Crags State Park.

Just east of the Klamath Region and also emerging from Oregon is the Cascade Region, which contains the Cascade Range as its sole range. The Cascades are part of a volcanic range that extends northward into Canada and Alaska. In California the Cascades are bounded by the Modoc Plateau on the east, the Klamath ranges on the west, and the Sierra Nevada on the south. The two major mountains in this chain within California, Mt. Shasta and Mt. Lassen, are still considered active volcanoes. Mt. Lassen erupted most recently between 1914 and 1917. Lassen Volcanic National Park contains strikingly beautiful reminders of the force of the eruption, in the form of bubbling mud pots, hot springs, and steam vents. Many orchids also occur here. One of our *Platanthera* hybrids, *P.* ×*lassenii,* was described from a specimen collected on Mt. Lassen.

The Sierra Region is coterminous with the Sierra Nevada, the longest continuous mountain range wholly contained within the United States.

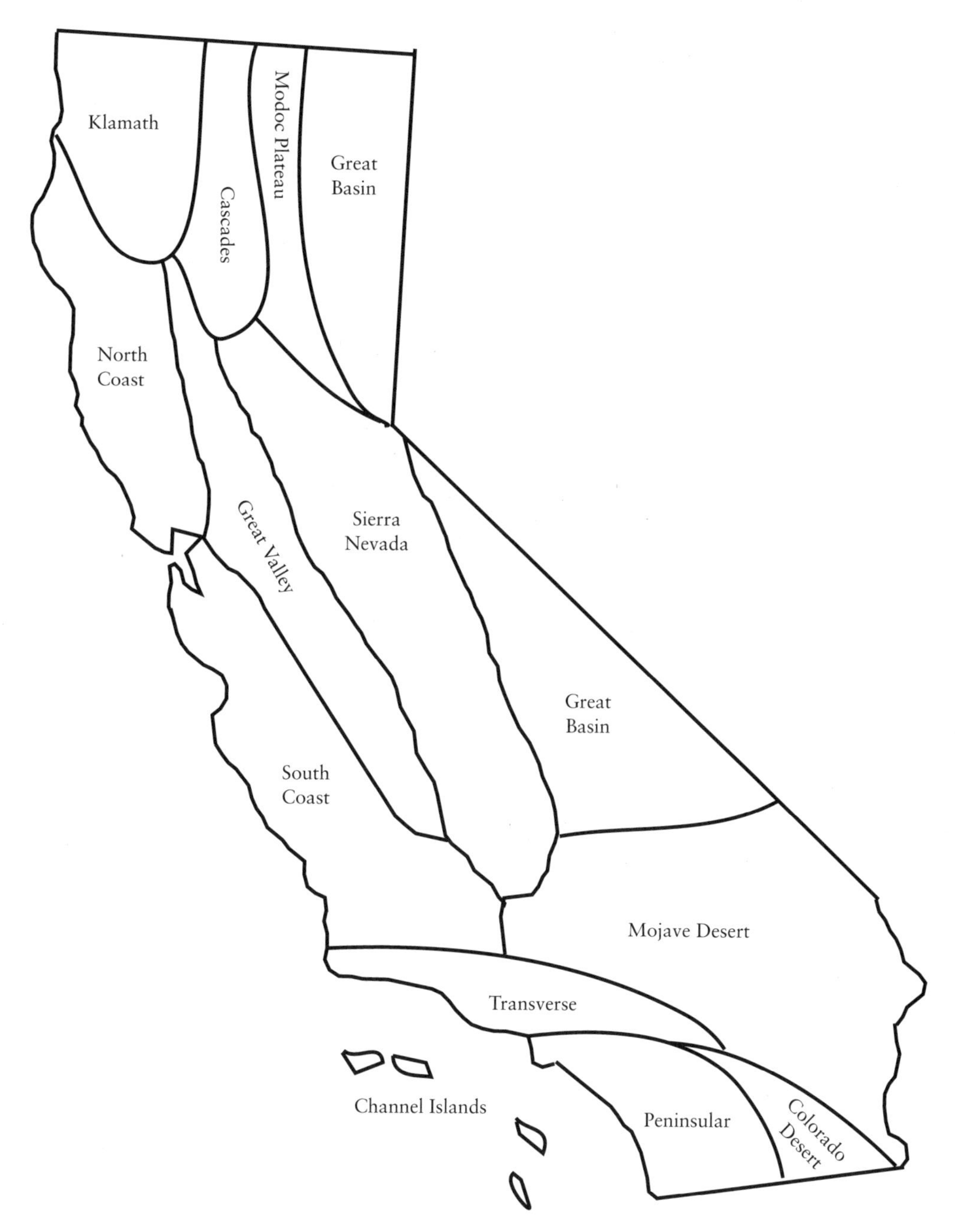

Map 3. Principal geographic regions of California

Mt. Whitney, in the southern portion of the range, is the highest peak in the lower 48 states. The eastern slope of the Sierra Nevada, much steeper and much less heavily forested than the western approach, is an example of extreme uplifting. Within the Sierra are many State and National Parks, including world-famous Yosemite National Park. Lesser known, but also of awe-inspiring beauty, are Sequoia National Park and Calaveras Big Trees State Park. Within all three of these parks are protected *Sequoiadendron giganteum,* the giant Sequoias, the largest living things. In the shadows of these giants grow many of our orchids, 15 in Yosemite alone. The northern half of the Sierra Nevada is richer in orchid species than the southern half. The lower limit for *Cypripedium montanum* is just south of Yosemite, and it is the only *Cypripedium* found that far south. Farther to the north, however, in Plumas County, all three of our *Cypripedium* grow. As might be expected because the western slope is wetter and more forested than is the eastern slope, it has a more diverse and more abundant orchid community.

The North Coast Region starts just south of the Klamath Region and ends at San Francisco Bay in Mt. Tamalpais State Park. Major ranges in this region are the King Range and the Mayacamas, South Fork, and South Yolla Bolly Mountains. This region is extremely rich in orchids and embraces the southern limits of *Listera caurina, L. cordata,* and *Corallorhiza mertensiana.* Contained within the North Coast ranges are some spectacular river valleys and monumental stands of *Sequoia sempervirens,* the coast redwood, the tallest trees in the world. The redwoods are taller than their mountain cousins the giant sequoias, but do not have as great a total mass because of their generally less massive trunk. These trees once grew in a nearly continuous band from Oregon to the Santa Lucia Mountains in Monterey County. Heavy logging has greatly reduced their numbers, but not their majesty, and magnificent groves are protected in several State and National Parks. One of our most beautiful orchids, *Calypso bulbosa,* grows in these groves, and has been given the local name of redwood orchid. Rainfall can be heavy in this region. In 1964 a series of severe winter storms pounded down on hillsides denuded by logging to produce a flood of almost biblical proportions. In Humboldt County, near Humboldt Redwoods State Park, reminders of nature's force are seen in the high-water marks of the 1964 flood, tacked to poles towering 6 meters above the roadbed.

The South Coast Region starts below San Francisco Bay and continues until terminated by the Transverse Ranges in Southern California. The South Coast Region includes about ten different ranges, but the Santa Cruz and Santa Lucia Mountains along the central coast support the largest number of orchid species. Within the Santa Cruz range lies the epicenter of the Loma Prieta earthquake that caused such severe damage around San Francisco Bay and well to the south, just prior to the 1989 World Series. Several orchids reach their southernmost coastal limits in these mountains. *Cephalanthera austiniae, Goodyera oblongifolia, Calypso bulbosa, Corallorhiza maculata,* and *C. striata* all have their southern coastal terminus here, although *C. maculata* and *C. austiniae* have both been collected inland at higher elevations in Southern California. *Piperia* reaches its peak here, and all but two members of that genus are found in multiple locations in the South Coast Region. *Piperia yadonii,* an orchid endemic to California, grows along the coast in the shadows of the Santa Lucia Mountains in Monterey County.

Most of the mountain ranges in California have a north-south orientation, but the mountains that dominate the Southern California counties of Santa Barbara, Ventura, Los Angeles, and San Bernardino, and parts of Kern County, have an east-west axis. These are called the Transverse Ranges. The major ranges in this region are the Santa Ynez, Santa Monica, San Gabriel, San Bernardino, and Tehachapi Mountains. Many peaks in these ranges exceed 2750 meters, providing Sierra-like habitats in Southern California. These southern ranges have a surprisingly diverse orchid flora. *Piperia michaelii* reaches its southern limit in these mountains, and *P. cooperi* reaches its northern limit here. Two disjunct orchid populations occur in the Transverse Region. *Malaxis monophyllos* var. *brachypoda* has been reported from only two places in California: the San Bernardino Mountains and the adjacent San Jacinto Mountains of the Peninsular Region. One of only two Southern California collections of *Cephalanthera austiniae* was made in the mountains of San Bernardino County. This species also grows nearby in the Peninsular Ranges in San Diego County, but its next nearest location is in Monterey County, along the coast far to the north. Some of the Transverse Region's mountains are surrounded by, or on the edge of, the most heavily populated area in California. The Santa Monica Mountains cut right through metropolitan Los Angeles, with the teeming San Fernando Valley, site of 1994's Northridge quake, on one side and Los Angeles on the other. Just to the north, the San Gabriel Mountains form the backdrop for Los Angeles, and the San Bernardino National Forest, covering parts of both the San Gabriel and San Bernardino ranges, is the most heavily visited National Forest in the United States. Because of the still expanding population and its insatiable appetite for land, much orchid habitat has been lost and more is being destroyed each year. Significant expanses of habitat have been protected, though, by various agencies, and these preserved areas provide places within the access of millions where we can still visit orchids in their natural conditions.

South of the Transverse Ranges, the Peninsular Region returns to the more familiar north-south orientation. The major ranges in this region include the Santa Ana, Santa Rosa, San Jacinto, Agua Tibia, and Laguna Ranges. The higher elevations of these ranges extend Sierra-type habitats into San Diego County, and support orchids not normally expected in this otherwise semi-arid climate. Mt. San Jacinto, Mt. Palomar, and Cuyamaca Peak all provide homes for *Corallorhiza maculata* and *Platanthera dilatata* var. *leucostachys*. *Epipactis gigantea* and some of the *Piperia* species inhabit the lower-elevation foothills. Orchid habitat in the rapidly expanding counties of Riverside and San Diego, like that in Los Angeles County, is falling to development at an increasing rate, although some prime areas are protected in State Parks, Wilderness Areas, and Nature Conservancy reserves.

The Basin Region, with at least 50 independently identified ranges, has the most mountain ranges of any of the state's eight regions. In the southern part of the state, most of the Basin ranges are desert and support few orchids. The White Mountains, the imposing range standing across the Owens Valley from the Sierras, is the most spectacular in the Basin Region. White Mountain Peak, at 4342 meters, is the third highest in California. In the upper elevations of this range grow the Bristlecone Pines, among the oldest living things. The age of the oldest of these twisted and gnarled trees, the Methuselah Tree, has been estimated at 4680 years. For all their bulk and majesty, the White Mountains contain only four orchid species: *Epipactis gigantea* and *Platanthera hyperborea* are readily found, but *Corallorhiza maculata* and *Spiranthes romanzoffiana* are known from only a few locations there. The Basin Region also enters the extreme northeast corner of the state, where the Warner Mountains are the dominant range. The Warner Mountains are much wetter than are the southern Basin ranges, and support a larger orchid population, including *Goodyera oblongifolia* and *Cypripedium montanum. Platanthera stricta* is more easily found in the Warner Mountains than in the other regions of California.

The California landscape is also defined by its valleys and deserts. The principal valley, known as the Central Valley, stretches from the Klamath Region to the Transverse Ranges, sandwiched between the mighty Sierra Nevada and the two Coast Ranges. This agricultural mother lode produces vast amounts of food for the nation's tables. The Central Valley is drained by two major rivers, the Sacramento and the San Joaquin. The northern part of the valley, through which the Sacramento River flows, from north to south, is sometimes called the Sacramento Valley. Likewise, the southern valley, drained by the San Joaquin River, flowing from south to north, is called the San Joaquin Valley. The two rivers merge in the Sacramento–San Joaquin Delta and, from there, empty into San Francisco Bay. Because it drains the wetter part of the state, the Sacramento River still flows year-round through the northern valley. The San Joaquin River, however, has been tapped so much by dams and reservoirs, to quench the thirst of agriculture and cities, that by the time it leaves the mountains it is a river in name only, regaining a semblance of its former glory only during the rainy season. Although a prime example of agricultural bounty, the Central Valley is sparsely populated with orchids. There are no records of any orchids in Kings, Sacramento, Solano, or Sutter Counties. The nearly ubiquitous *Epipactis gigantea,* however, must be in these counties somewhere, and it is probably just a matter of time until it is found.

Another major valley in California is the Owens Valley, located between the eastern edge of the Sierras and the White Mountains. The Owens River once flowed through this beautiful valley, feeding many lakes along its route. Early this century, however, the burgeoning city of Los Angeles secured the water rights to much of the watershed for the river, and most of the water that once flowed through the river now flows directly to Los Angeles via aqueducts. The river itself and its lakes have largely dried up, but the drive through Owens Valley remains truly outstanding, with the towering Sierra Nevada on one side and the equally majestic White Mountains on the other. Along the way, Highway 395 meanders around now-dormant cinder cones and through ancient lava flows. Unlike the Central Valley on the western side of the Sierras, the Owens Valley does support some orchids. *Epipactis gigantea* grows in several locations, and *Spiranthes porrifolia* grows on alkali flats in the northern part of the valley.

The deserts are yet another significant feature of the California landscape. Most commonly, we think of the two major desert regions, the Mojave and the Colorado, in the southeast quadrant of the state when we envision the deserts, but in the strictest sense even the White Mountains are desert, because their location in the rain shadow of the Sierras results in low precipitation levels. Death Valley National Monument in the Basin Region is the lowest and the hottest spot in the United States. Anza-Borrego Desert State Park and Joshua Tree National Monument preserve other important expanses of desert. Normally, we would not think to go to the desert to look for orchids, but all of the deserts, including Death Valley, support multiple colonies of *Epipactis gigantea.* This remarkable orchid survives in alkaline soils near a constant source of moisture, such as is found at desert springs and oases.

Habitat Protection in California

California's ever-growing population demands more and more land each year for homes and businesses and other human concerns. Inevitably, habitat that once supported not only orchids, but also other rare and beautiful plants and animals, is lost each year to the relentless forces of development. But for more than a hundred years, fortunately, and still today, farsighted individuals and groups have recognized the uniqueness and beauty of this land and have fought to preserve it. Approximately half of the land in the state is in public ownership, most of it within National

Forest. But because National Forest lands can be and are used for economic purposes, such as logging and mining, the loss of orchid habitat occurs on public as well as private lands. Some lands within the National Forest are off limits to development, and large areas are totally protected as Wilderness Areas, Research Natural Areas, or other special study or resource areas where even vehicle access is prohibited.

California's State Park System provides protection to orchid habitat through the nearly 300 units under its control. These units range from vast acreage, in parks such as Calaveras Big Trees State Park, Anza-Borrego Desert State Park, and multiple reserves and Wilderness Areas, to small plots, in various State Beaches and Historic Parks. All of these types of units potentially protect some orchids, often many species.

California also contains several units of the National Park System. In places like Yosemite National Park, Sequoia National Park, Redwood National Park, the Point Reyes National Seashore, the Santa Monica Mountains National Recreation Area, and Death Valley National Monument, orchid habitat is protected in its natural state by the National Park Service.

City and county parks are very valuable resources, and many protect considerable additional habitat. Private reserves are another method of protecting plant communities and habitat. Two such reserve systems operating on the national and international level are protecting significant acreage in California. The Nature Conservancy (see Holing, 1988) maintains over 20 preserves in California. Many are open to the public during certain hours, or can be entered by permission. The second of these systems, managed by the Audubon Society, also maintains a group of private reserves in California. Some conservancy agencies have been set up to protect a local interest; the Santa Monica Mountains Conservancy, for example, specializes in land preservation in the Santa Monica Mountains, in Ventura and Los Angeles Counties.

None of these park or reserve units, whether in the local, state, national, or private systems, was set aside specifically to preserve orchid habitat or plants. Some were established to protect unique plant communities or habitat areas. Others were created to protect natural formations or other areas of special beauty. Whatever the reason for their creation, by preserving wild areas and natural environments, they also preserve some of the habitat of the wild orchids in California. Many have the advantage of prepared and maintained trails, making for easy access. Entrance permits, if required, are usually readily obtainable, and permission to enter private property, necessary elsewhere, is not an issue. Most important, because of the vast total acreage in these multiple systems, and their broad distribution throughout the state, every one of California's wild orchids is in protected habitat in at least part of its range.

Key to the Genera of Wild Orchids in California

The key below assumes that the reader has found a plant in bloom, has identified it as an orchid, and desires to know its genus. Once that is established, the species it represents (within that genus) can then be determined by referring to the key included with the discussion of the indicated genus. Botanical keys are often difficult for an amateur (and some professionals) to follow, and the keys in this book are no exception. In part because of the variability inherent in orchids, identification using keys is not always successful, particularly with *Piperia, Platanthera,* and *Spiranthes. Piperia* was separated from *Platanthera* partly because its leaves are usually faded at anthesis, and the key makes use of this fact. Infrequently, but often enough to be confusing, a piperia will flower with still-green leaves. In both *Platanthera* and *Spiranthes,* hybrids and intermediate forms will frustrate anyone trying to identify them using keys. The keys should thus be used to *complement* the species descriptions, the drawings, the maps, and the color plates, not as the sole means of identification. The usefulness of the maps to aid identification should not be underemphasized. For example, if you are in the coastal mountains of Southern California, the plant you are studying is most likely not a *Cypripedium.* Likewise, if you are in Del Norte County, the *Piperia* you wish to identify is most likely not *P. cooperi.*

1. Flowers with two perfect anthers — *Cypripedium*
1. Flowers with one perfect anther:
    2. Plants without green leaves at any growth stage:
        3. Lip without an epichile — *Corallorhiza*
        3. Lip with an epichile:
            4. Epichile ridges or protuberances less than three; plants and flowers pink or white; sepals and petals usually spreading — *Epipactis*

4. Epichile ridges or protuberances more than three; plants and flowers white except for yellow marks on lip; sepals and petals usually not spreading *Cephalanthera*

2. Plants with green leaves at least prior to anthesis:
   5. Leaves basal:
      6. Leaf single:
         7. Flower single, the lip pouchlike, the pouch with bifurcated apex *Calypso*
         7. Flowers multiple, tiny, the lip broadly triangular *Malaxis*
      6. Leaves two or more:
         8. Leaves arranged in a rosette *Goodyera*
         8. Leaves not in a rosette:
            9. Flowers arranged in spirals around inflorescence axis; flowers without spurs *Spiranthes*
            9. Flowers not arranged in clear spirals; flowers with spurs *Piperia*
   5. Leaves cauline:
      10. Leaves two, opposite or subopposite, near middle of stem *Listera*
      10. Leaves many, alternate, scattered along stem:
          11. Leaves smooth; flowers spurred *Platanthera*
          11. Leaves plicate; flowers not spurred *Epipactis*

2. *Calypso* Salisbury

Paradisus Londinensis: pl. 89. 1807.
Etymology: *Calypso* is from the Greek name of the sea nymph in Homer's *Odyssey.*

Calypso (ka - lip' - soe) is a monotypic, circumpolar genus with four recognized varieties of its single species: *C. bulbosa* var. *bulbosa,* found in Europe and Asia; *C. bulbosa* var. *speciosa,* found in Japan; and *C. bulbosa* var. *americana* and var. *occidentalis,* both found in North America. Because the lip forms a pouch, or slipper, Linnaeus originally included calypso with *Cypripedium.* Subsequently, Salisbury determined that the structural differences in the lip and column merited the erection of a separate genus.

The two American varieties of *C. bulbosa* are distinguished by their geographic distribution and the characters of their flowers. Some authorities have argued that they are distinct species. *Calypso bulbosa* var. *americana,* the eastern race, has bright-yellow markings and hairs at the opening to the pouch; *C. bulbosa* var. *occidentalis,* the western race, has white throat markings and hairs, and its hairs are fewer. The two varieties overlap only in British Columbia (see Long, 1980). Wilken and Jennings (in Hickman, ed., 1993) do not recognize the two varieties of *C. bulbosa,* but the differences in the plants are so pronounced that the varietal distinction should be maintained.

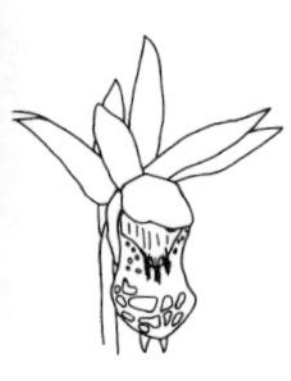

Calypso bulbosa (Linnaeus) Oakes var. *occidentalis* (Holzinger) Calder and Taylor

Canadian Journal of Botany 43: 1393. 1965.

Etymology: The specific epithet, from Latin *bulbus,* refers to the bulblike nature of the corms.

Synonymy:
Calypso occidentalis (Holzinger) A. A. Heller, Bulletin Torrey Botanical Club 25: 193. 1898.
Cytherea occidentalis (Holzinger) Heller, Muhlenbergia 1: 132. 1906.
Cytherea bulbosa (Linnaeus) House var. *occidentalis* (Holzinger) Cockerell, Torreya 16: 185. 1916.

Common names: calypso, fairy slipper, redwood orchid, Venus slipper, deer head orchid.

Plate 1

Description: Plants and flowers of *C. bulbosa* (bul - boe′ - sa) var. *occidentalis* (ok - si - den- tah′ - lis) show considerable variation in size, even in plants within a few meters of each other. Normally, calypso blooms on an inflorescence 15 to 20 cm tall, although inflorescences under 10 cm are common, and larger plants extend up to 32 cm from ground level to the uppermost tip of the flower. The solitary oval, basal leaf typically measures approximately 6 × 5 cm, but on larger plants can be as much as 11 × 8 cm. The leaf is entirely separate from the inflorescence, and because the base of its blade barely clears the humus, appears almost stemless. The shallowly rooted plants grow in humus and forest duff, rarely penetrating the soil. The corms have very few roots, and the roots, short and slender, sometimes develop coralloid masses. On blooming plants it is often possible to see the top of the corm. Typically, each corm produces only a single leaf and a single flower, but multi-flowered specimens, with as many as four flowers per corm, each flower on a separate stem, occur infrequently.

The natural spread of the flowers varies between 2 and 5 cm, and the purple stems seem too dainty to hold the blossoms. The bract that protects the bud as the stem is elongating is also purple, looking almost like an extension of the flower. The lightly veined lanceolate sepals and petals are nearly identical and are held like a crown above the column. The lip, a complete pouch, gives rise to the common name fairy slipper. The top of the lip, or lamina, is covered with purplish spots, and extends beyond the sides of the pouch like a skirt, hiding it almost completely, though the bifurcate tip of the pouch extends beyond the edge of the skirt. On some plants the spots on the lip are tiny and distinct, while on others (never on the same plant) they merge into large blotches. The area near the throat of the pouch is white, and is decorated with three ridges covered with white hairs. The hairs at the opening to the pouch in var. *occidentalis* are fewer, straighter, and more slender than those in var. *americana* (see Heller, 1898). Purple to reddish veins line the inside rear of the pouch. The column, shaped like an inverted saucer with wide side lobes, extends over the orifice of the pouch. Parsons (1907, p. 216) called the column "a curving concave petal in the shape of a hollow hemisphere." Calypso has a distinct, pleasant aroma of highly variable intensity. The fragrance of some plants is detectable several meters away, while that of others is barely noticeable. The capsules are ellipsoid, and as they mature, the stem straightens from its nodding posture to hold them erect.

Flowers of *Calypso bulbosa* var. *occidentalis* exhibit considerable plant-to-plant color variation, ranging from rich purple through shades of pink to white. Faint orange highlights appear on the lip of some flowers, and a small percentage of flowers are salmon-colored. White flowers, which occur infrequently, usually retain some brownish coloration in the spots on the lip, but lack purple pigmentation in the flower stem and bract. In Trinity County there are two widely separated groups of plants with nearly pure-white flowers. In one group the flowers are white except for yellow highlights on the lip and yellow markings on the inside of the pouch. In the other group the flowers are milk-white with only a slight hint of yellow on the lip, near the hairs. In both cases the flower stems are green instead of the normal purple, and the upper bracts are the same white as the sepals and petals. The leaves and stems of these plants show no traces of red pigment. Both white-flowered groups bloom amid hundreds of normally colored flowers. White flowers also occur in var. *americana,* but rarely; that white form has been named *C. bulbosa* var. *americana* f. *candida* (see Whitling and Catling, 1986).

The bloom cycle of var. *occidentalis* differs considerably from those of our other orchids. In the fall, as early as September, each corm sprouts a single, dark-green, oval leaf that lasts through the winter, surviving under the snow in cold areas. The usually solitary flower appears at the first signs of spring and is able to withstand late frosts. Shortly after the flowers bloom, the leaf fades for the summer. By June in most areas of California, it is impossible to tell that calypso ever bloomed, for even the remnants of the seed capsule have disappeared. *Calypso* is one of only three deciduous North American orchid (the others, neither of them represented in California, are *Tipularia* and *Aplectrum*) genera whose plants have winter leaves.

The pollinators of *C. bulbosa* var. *occidentalis* are attracted by general food mimicry rather than a specific reward (see Ackerman, 1981). The bifurcate structures at the end of the pouch that might be assumed to be nectaries are not. Ackerman found no traces of nectar in the flowers, and he identified the pollinators as bees of the genera *Bombus* and *Psithyrus.* After landing on the lamina of the lip, the bee enters the pouch searching for the nonexistent food. While exiting the pouch, the bee brushes against the column overhanging the opening to the pouch, and the pollen is deposited on the bee. The pollen granules are then transferred to the next flower visited. Ackerman (1981) reports fairly low fruit set, varying between 0 and 34%. The hairs on the lip may simulate pollen-bearing anthers, thereby attracting bees to the nectarless flowers (see Wood, 1986).

Distribution: *Calypso bulbosa* var. *occidentalis* grows from California to the southern tip of Alaska and eastward to Idaho. Within California, it occurs in only 11 counties in the northern and central part of the state. Its southernmost location is in Santa Cruz County, and except for San Francisco, it ranges continuously up the coast to Del Norte County. It also ventures away from the coast into the mountainous regions of Siskiyou,

Humboldt, and Shasta Counties. In Napa County, calypso is carried on the plant list of a State Park, and on the list of a local flora, so it is fairly certain to occur there. Records of calypso in the Santa Cruz Mountains counties of Santa Cruz and San Mateo are relatively recent. The first reference of calypso south of Marin County is to a 1963 discovery from Big Basin State Park in Santa Cruz County (see Crandall, 1963), which corresponds to the earliest herbarium record from the Santa Cruz Mountains. Did the beautiful calypso escape notice until then, or was it simply undocumented? Is calypso perhaps expanding its range? These questions may be unanswerable, but they add another challenge to the search for calypso.

Habitat: *Calypso bulbosa* var. *occidentalis* grows from sea level to nearly 1800 meters, although the plants are much more numerous below 600 meters. Calypso prefers the dry forest floor, under firs, pines, spruce, redwoods, or oaks. Beneath the forest canopy it grows in the open or seeks the additional protection of low-growing shrubs, poison oak, or grasses. It favors partly shaded conditions, though it will grow in the dense shade of redwood forest. Calypso is more numerous on flat or nearly flat terrain, but it also grows on slopes and even on fairly steep hillsides. Along the coast, calypso blooms in pine and spruce forest at the edges of sand dunes. In the damper portions of its habitat, it sometimes grows in moss, or colonizes and blooms atop rotting logs and tree stumps.

Blooming season: The blooming season for calypso lasts from late winter to summer. The coastal flowers open first, appearing at the beginning of March. Individual flowers last about one month, and newly opening plants extend the season in each area to about two months. By the end of May most flowers below 600 meters have faded. The start of blooming is delayed at the higher elevations, where flowering typically lasts into June and occasionally until mid-July. The best displays, however, are in April and early May.

The beauty of the individual flowers is multiplied when blooming occurs in great masses, as it often does. The massed flowers are often visible to motorists passing by. Several acres of predominantly fir-covered hillside at one Trinity County location are practically painted with *C. bulbosa* in the spring. As far as one can see up the hill, and across several hundred meters in breadth, little bursts of purple dot the forest floor. In 1991, blooming plants at this location numbered in the thousands, with an equal number of younger plants. Near the peak of blooming, plants in all stages will be present. The most advanced ones will have already formed capsules; others will still be in bud, with some buds just emerging from the ground. The flowers, seedlings, and young plants are so dense in spots that each step must be chosen carefully in order not to tread on them or damage them.

The forests of Northern California are home to many orchids besides calypso, and several of them overlap their blooming seasons, especially in May. *Corallorhiza maculata, C. mertensiana,* and *C. striata* bloom near calypso, as does *Cypripedium fasciculatum. Listera cordata* and calypso will sometimes intermingle so densely that it is hard to photograph either plant without fear of damaging one or the other. The leaves of *Goodyera oblongifolia* grow within sight of the fairy slippers, but will not even show spikes until calypso is out of bloom. The *Piperia* that will be blooming in late summer may just be starting to put up an inflorescence when calypso is in its prime.

Conservation: Timber is the basis for a major industry in Northern California, and logging operations are the leading cause of habitat destruction for calypso. Fortunately, much prime habitat is protected within parks and preserves. Efforts to save more redwood forest, and hence more calypso territory, are continuing, by both the parks and groups such as the Save-the-Redwoods-League. In many places calypso has reestablished itself in second-growth redwood forest originally logged 100 or more years ago. Besides logging there are two major threats to calypso. The shallow growth habit of these plants makes them easy prey for foraging feral pigs, and in some areas the pigs have destroyed large colonies. The other major threat is people, who can be destructive both by accident and by deliberate action. Some popular colonies are visited by hundreds each year, and in their haste to see and photograph the flowers, the visitors sometimes step on and uproot nearby plants. The beauty of calypso makes it a great target for collectors, and more are lost by deliberate removal than by accident. In one of the two white-flowered groups in Trinity County, all of the plants were removed by collectors. Visits to the same area in three subsequent years failed to discover any remaining or new plants, which suggests that the colony was destroyed.

Notes and comments: For several reasons, Calypso has a special meaning for me. It officially marks the beginning of the orchid blooming season for my family. Technically, calypso is our third orchid species to open, behind *Corallorhiza maculata* and *C. striata,* but in mid-March each year my first trip into Northern California is to see calypso. It also is special because I looked for it for 11 years before finding it. For the first 10 years, I waited until June to go searching, and always missed the blooming season. Even now, with 10 years more experience, finding it in bloom in June takes considerable work and a good deal of luck. It is much easier to go earlier in the spring and see calypso in peak bloom.

3. *Cephalanthera* L. C. Richard

De Orchideis Europæis: 29. 1817; Mémoires du Muséum d'Histoire Naturelle Paris 4: 51. 1818.

Etymology: *Cephalanthera* derives from the Greek words for head and anther, in reference to the relative positions of the anther and column.

Cephalanthera (sef - al - an - thair′ - a) consists of some 15 species that grow primarily in the Northern Hemisphere. *Cephalanthera* is closely related to *Epipactis,* another genus present in California. The primary differences are in the flower stem and column structure. The flowers of *Cephalanthera* are stemless, whereas those of *Epipactis* have a clearly defined stem. The column of *Cephalanthera* is considered very primitive (Davies et al., 1988), with a simple, sticky stigma and no rostellum. Visiting insects, probing for nectar, brush against the stigma and come away bearing some of the sticky substance. As they back out of the flower, parts or all of the two pollinia adhere to the sticky substance, to be deposited, perhaps, on the next orchid visited. *Cephalanthera austiniae,* the only member of the genus native to the United States, is mycotrophic.

Mycotrophic plants derive food from a mycorrhizal relationship with a fungus. A compatible fungus invades the root mass of the orchid, and the orchid uses food material already digested by the fungus. Although most, if not all, orchids need a mycorrhizal fungus to germinate their seeds, mycotrophic orchids never outgrow that need; if the fungus dies for any reason, the orchid also dies. Often *C. austiniae* and other orchids that largely lack chlorophyll are incorrectly referred to as saprophytes. The term *saprophyte* as defined by Dressler (1981) describes a plant that "does not manufacture its own food by photosynthesis, but uses organic material (previously manufactured by other plants) in its substrate." Dressler adds that vascular plants (all so-called higher plants) are incapable of digesting organic material in the leaf litter, and the term "saprophyte" technically does not apply to them. He prefers the term "mycotrophic." MacDougal (1899) suggested the use of the term "symbiotic saprophyte," but most recent authors, like Dressler, use "mycotrophic." Other mycotrophic orchids in California are in the genera *Corallorhiza* and *Epipactis*.

Cephalanthera austiniae (A. Gray) A. A. Heller

Catalogue of North American Plants, ed. 2: 4. 1900.

Etymology: The specific epithet honors Rebecca Merritt Austin, who collected the type specimen.

Synonymy:
Chloraea austiniae Gray, Proceedings American Academy of Arts and Sciences 12: 83. 1876.
Cephalanthera oregana Reichenbach, Linnaea 41: 53. 1877.
Epipactis austiniae (A. Gray) Wettstein, Österreichische Botanische Zeitschrift 39: 429. 1889.
Eburophyton austiniae (A. Gray) Heller, Muhlenbergia 1: 49. 1904.
Serapias austiniae (A. Gray) A. A. Eaton, Proceedings Biological Society of Washington 21: 66. 1908.

Common names: phantom orchid, snow orchid.

Plate 2

Description: One look at *Cephalanthera austiniae* (os - tin′ - ee - eye) convinces you that it deserves the common name phantom orchid. The pure-white stem can exceed 38 cm in height and bear more than 25 flowers, each about 2 cm across. Sometimes the bracts on the stem are very leaflike, reaching nearly 8 cm in length, but because the plant is mycotrophic (see the genus description), it bears no leaves. The flowers are white, except for yellow markings on the lip, and have a scent reminiscent of vanilla. The sepals are elliptic-lanceolate, and the upper petals are elliptic-oblong. The three-lobed lip is hinged near its middle. The epichile, sharply recurved, has five raised ridges on its middle lobe. The hypochile is saccate, with three rows of bumps and a yellow dot at the bottom of the sack. The flowers do not open fully, retaining a slightly cupped appearance. Perhaps because they are white, the flowers seem prone to damage, and usually some on each plant abort without normal opening. The plants have a slender rhizome with multiple thick, fibrous roots. The obovoid-ellipsoid capsules are held erect, and usually darken quickly from the pure white they exhibit when the flower is fresh.

The phantom orchid was first named *Chloraea austiniae* by Gray (1876). In a nearly simultaneous publication Reichenbach (1877) called it *Cephalanthera oregana.* Heller (1900) recognized Reichenbach's work and placed the species in the genus *Cephalanthera,* retaining Gray's specific epithet. Heller (1904) subsequently had second thoughts and created a monotypic genus, renaming the plant *Eburophyton austiniae,* a name later adopted by many authors. He observed that *Cephalanthera austiniae* showed resemblance to both the South American genus *Chloraea* and the circumtemperate genus *Epipactis.* He justified creating a new genus for it on the basis of its mycotrophic growth habit, saying, "In my opinion a plant of such peculiar habit should be separated on that point alone, even if floral and fruit characters are similar to some other plants." Heller's assumption, however, was incorrect; other members of the genus *Cephalanthera* are also mycotrophic (Catling and Sheviak, 1993). Catling and Sheviak place the phantom orchid in the genus *Cephalanthera* because of floral morphology. They also observe that the correct spelling of the specific epithet is *austiniae,* not the more commonly used *austinae.*

Distribution: *Cephalanthera austiniae* is confined to a narrow band in California, Oregon, Washington, and British Columbia. Personal observations and herbarium records place it within 28 counties in California, and a flora maintained by the local chapter of the California Native Plant Society places it in Napa County as well. It is most common from Monterey County north along the coast, and from Tulare County north in the Sierras. The type specimen of *C. austiniae* was collected near Quincy in Plumas County by Rebecca Merritt Austin, an indefatigable botanist from the turn of the century, and Gray named it in her honor. Native orchids collected by Mrs. Austin are in many of the herbarium collections in California.

There are only two records of the phantom orchid from Southern California. The first is from San Bernardino County, where John Roos collected it in 1934. Although still a teenager, he recognized the plant as unusual for the area, and his father took a sample of it to Philip Munz, then working at Pomona College. Munz identified it as C. *austiniae* and subsequently documented the find in his *Manual of Southern California Botany* (Munz, 1935), and later in his *Flora of California* (Munz, 1968). Roos visited the plants intermittently between 1934 and 1941, but in 1992 he was unable to locate the phantom orchid near his original collection site. The second sighting of the species in Southern California was made by Geoffrey Levin of the San Diego Natural History Museum, who discovered a small colony in San Diego County in July 1992. This recent discovery in a heavily visited area offers hope that C. *austiniae* is still hiding somewhere in San Bernardino County and other undiscovered haunts statewide.

Habitat: *Cephalanthera austiniae* grows between sea level and 2200 meters elevation. Its favorite habitat is open but shaded areas near or just above the banks of small streams. It frequently inhabits wooded bottomland near rivers, or between streams, and another favored habitat is dry, wooded slopes. It will also grow under bushes and in nearly full sun at the edges of clearings. Some colonies grow in long, straight rows as if they were growing on top of a buried decaying tree. Along the coast it inhabits forests at the edges of beach sand dunes.

Blooming season: The phantom orchid is one of our earliest bloomers, first opening in early March near sea level and in the mountains of Monterey County. At the upper reaches of its elevation range it may still be in bloom near the first of August. Within each area of its occurrence, the numbers of C. *austiniae* vary greatly from year to year. In 1988 a very large colony bloomed in the Trinity Mountains of Siskiyou County. It consisted of many hundreds of plants scattered thickly over several hundred square meters, even continuing on both sides of a road, but in 1989 fewer than a dozen plants bloomed at that site. Both Petrie (1981) and Long (1979) have commented on this species' year-to-year fluctuations in numbers and have observed that the phantom orchid can lie dormant underground for years before reappearing. Long quotes data from a Mrs.

Tye documenting variation in blooming over a 15-year period. The number of plants sometimes increased over 100% from one year to the next, and year-to-year decreases were also great. One year no plants came up, but some did reappear the following year.

Cephalanthera austiniae often grows in the same area as several other orchids. It blooms at the same time and often within a few meters of *Corallorhiza maculata, Cypripedium fasciculatum, Listera convallarioides, Piperia transversa, P. unalascensis,* and *Platanthera dilatata.* Frequently growing nearby, but either already out of bloom or not yet flowering, are *Calypso bulbosa, Cypripedium montanum,* and *Goodyera oblongifolia.*

Conservation: The phantom orchid is considered by some to be rare in parts of its range. According to Haskin (1967, p. 65), "It is one of the rarest of our flowers" and "is no where abundant." Scheffer (1970, p. 31) observed that it is "one of the rarest flowers in the west," and Long (1979, p. 30) noted that it "is among the rarest of plants in the whole of Canada." In California it is neither rare nor endangered, and is locally common in several counties. Major portions of its range are safely within the confines of State Parks and National Parks. It occasionally occurs even in campgrounds, and because of its white color is easily spotted from roads and highways. In parts of its range in California, however, its habitat is threatened by logging and development. The site of Roos's discovery in southern California has been heavily developed, which is probably why recent attempts to locate it there have not been successful.

4. *Corallorhiza* (Haller) Chatelain

Spec. Inaug. de Corallorhiza: 6. 1760.

Etymology: *Corallorhiza* derives from the Greek words for coral and root, in reference to the resemblance of the rhizome to coral.

Corallorhiza (ko - ral - oh - rye′ - za) is a mycotrophic orchid genus of about 12 species widely distributed in North America, as well as several species in Central America and one in Europe. Six species grow in the United States and Canada, and four of those occur in California.

Most of the *Corallorhiza* (the coral-roots) lack significant amounts of chlorophyll, and therefore lack green coloration. With no need to support photosynthesis, the residual leaves are reduced to mere bracts on the flower stem, and the plants are essentially rhizome, roots, stem, and flowers. The coral-roots are very colorful, exhibiting many shades of browns, reds, and yellows. The structure of the roots varies somewhat; those in some members of the genus resemble coral, hence the plants' common name. The coral-like roots of *Corallorhiza* are one of the main characters used to distinguish it from the very similar *Hexalectris* (which does not occur in California). Other differences include number of pollinia (four in *Corallorhiza,* eight in *Hexalectris*) and structure of the base of the lateral sepals (in *Hexalectris* the sepals are free at the base; in *Corallorhiza* the lateral sepals are united, usually forming a mentum). The four species in California differ widely in distribution and frequency of occurrence. *Corallorhiza maculata* is one of the most numerous and most widely encountered of our wild orchids, but *C. trifida* var. *verna* is one of the rarest orchids in California, as well as one of the most difficult to find.

Corallorhiza striata and *C. mertensiana* lie between these extremes in both distribution and ease of locating.

Key to the California Species of *Corallorhiza*

1. Lip three-lobed; mentum minute to well-defined; lip, sepals, and petals without stripes, but lip may have spots or blotches:
   2. Plants small, usually under 15 cm tall; lip less than 5 mm long — *C. trifida*
   2. Plants taller than 15 cm; lip more than 5 mm long:
      3. Column longer than 5 mm; mentum usually with a free tip, the lateral lobes of the lip not well defined or absent — *C. mertensiana*
      3. Column less than 5 mm long; mentum usually totally attached to ovary, the lateral lobes of the lip prominent — *C. maculata*
1. Lip entire, without lateral lobes; mentum absent; lip, sepals, and petals profusely striped, but lip without spots — *C. striata*

Corallorhiza maculata (Rafinesque) Rafinesque

American Monthly Magazine and Critical Review 2: 119. 1817.

Etymology: the epithet *maculata* is from the Latin word for spotted or stained.

Synonymy:
Cladorhiza maculata Rafinesque, American Monthly Magazine and Critical Review 1: 429. 1817.
Corallorhiza multiflora Nuttall, Journal Academy of Natural Sciences, Philadelphia 3: 138. 1823.
Corallorhiza mexicana Lindley, Genera and Species of Orchidaceous Plants: 534. 1840.
Neottia multiflora (Nuttall) Kuntze, Reviso Generum Plantarum 2: 674. 1891.
Corallorhiza grabhamii Cockerell, Torreya 3: 140. 1903.

Common names: spotted coral-root, large coral-root, many flowered coral-root.

Plates 3 and 4

Description: Typical plants of *Corallorhiza maculata* (mahk - yoo - lah′ - ta) are 40–50 cm tall, but some are as tall as 80 cm and others bloom at as little as 10 cm. Most spotted coral-roots bear 20–30 flowers, and robust plants occasionally produce over 40 flowers on a single stem. The natural spread across the sepals ranges from less than 1 to nearly 3 cm. The common name spotted coral-root derives from the white, three-lobed lip, which is usually dotted with few to many reddish or purplish spots. The oblanceolate sepals and petals and the short, curved column are also often spotted. On the central lobe of the lip are two prominent callosities. The shape of the lip is quite variable; typically, it spreads widely at the apex, with markedly undulate margins, whereas in other plants it is narrow and linear, with only a slightly undulate, upturned margin at the apex. Lips intermediate between the wide and narrow shapes are common, and plants with many different lip shapes bloom simultaneously in mixed colonies. The lateral lobes of the lip also vary: on most plants they are prominent, deeply lobed, and forward-pointing; on others the lobes are significantly reduced. The bases of the sepals extend backward along the ovary, forming a chinlike projection called a *mentum*. Plants with a well-defined mentum have been referred to as *C. masculata* Rafinesque var. *occidentalis* (Lindley) Cockerell. The capsules are ellipsoid and pendant.

Plants commonly recognized as *C. maculata* may in fact represent two distinct taxa, one constituting an early-flowering entity, the other a late-flowering entity (see Freudenstein and Bailey, 1987). Freudenstein and Bailey differentiate between the two by the size of the flower parts, particularly the lip, and the timing of the blooming season. They find the differences more pronounced in eastern plants, and believe that the characteristics tend to merge in western populations, which is consistent with my observations.

The flowers of the spotted coral-root vary in size and color, even within the same growing area. The sepals, petals, and stem exhibit many shades of red, brown, yellow, or purple. This penchant for variability has resulted in multiple varietal names that have been questioned by some authorities. Correll (1978), for example, believed that even the more striking variants were not worthy of taxonomic consideration.

The primary color variants encountered in the literature, all with variety status, are *flavida, intermedia, punicia, fusca,* and *immaculata.* Flowers within California correspond to the published descriptions of each of these. *Corallorhiza maculata* Rafinesque var. *flavida* (Peck) Cockerell represents lemon-yellow plants with pure-white, unspotted lips. Because of its color, *C. maculata* var. *flavida* is sometimes referred to as the yellow coral-root. Several other names have been used to describe the yellow coral-root. *Corallorhiza ochroleuca* Rydberg describes a pale-yellow flower with pure-white lip from Nebraska and Colorado. *Cor-*

allorhiza multiflora var. *sulphurea* Suksdorf also describes a pale-yellow form. Yellow plants and flowers occur sporadically throughout the range of *C. maculata* in California. In places only a single yellow plant will be mixed in with brown plants, whereas in other areas the yellow coral-root becomes dominant, far outnumbering the brown plants. The yellow coral-root is most easily found in the northern coastal mountains and in the Sierra Nevada. A variant with pure-yellow flowers and spotted lips occurs in portions of the range.

Occasionally, *C. maculata* Rafinesque var. *immaculata* Peck is used to refer to normally colored flowers with white, unspotted lips, but flowers with white, unspotted lips are occasional in most areas of the range, and in some locations, the white-lipped flowers are the majority, or only, form present.

Corallorhiza maculata var. *punicia* Bartlett identifies plants with bright reddish-purple stems and pure-white or brightly spotted lips. The bright coloration of the stem and bracts extends to the back side of the sepals and petals, but the front sides are much paler. Coral-roots meeting this description grow in many areas of California.

Corallorhiza maculata Rafinesque var. *fusca* Bartlett is "light cinnamon-drab," with sheaths of "Van Dyke brown" (see Bartlett, 1922). Light cinnamon-drab aptly describes the color of the most common spotted coral-root flowers in California. *Corallorhiza maculata* Rafinesque var. *intermedia* Farwell is, as the name implies, intermediate in color between *C. maculata* var. *punicia* and *C. maculata* var. *flavida.* It is most easily described as a purplish yellow with spotted lips. Plants matching this description are very widely distributed, and may appear with any of the other color types. Bartlett (1925) believed var. *intermedia* and var. *fusca* to be distinct varieties, but Farwell (1923) considered them synonyms. These are perhaps the most poorly defined of the *C. maculata* color forms. Like the others, they are of variable shades and intensity, and illustrate that there is too much variation in the color of *C. maculata* to give much meaning to any of the many varietal names.

The pollination of *C. maculata* has not been thoroughly documented. Luer (1975) published a picture of an *Andrena* bee carrying away pollen from the spotted coral-root. Kipping (1971) reported pollination by predaceous dance flies (Empididae), with the pollinarium becoming attached to the thorax. Bumblebees visit the flowers, but they have not been documented as pollinators. The percentage of fruit set is very high, nearing 100% in most areas and in most years.

Distribution: *Corallorhiza maculata* ranges widely across North America and is the most common coral-root in California, occurring in 41 counties. It grows along the length of the Sierra Nevada and in several places in Southern California, including localities in the San Bernardino, San Jacinto, and Cuyamaca Mountains. The spotted coral-root is one of only a few orchids, and the only coral-root, found in the White Mountains, on the eastern edge of Owens Valley. It also grows in the coast ranges as far south as the Santa Lucia Mountains. Because of habitat similarities, one may infer that the spotted coral-root also occurs in Alpine County, but it has not been confirmed there.

Habitat: *Corallorhiza maculata* inhabits dry, open forest, making its home in the litter of conifers and oaks between sea level and 2740 meters. In some dense timber stands it is the only herbaceous plant, and where conditions are suitable, large colonies develop. The spotted coral-root shows some tolerance for varied growing conditions, often blooming in the moist environment of creek banks or river bottoms in competition with other herbaceous plants. In those moist conditions it sometimes hides under low-growing shrubs. The spotted coral-root is one of only two wild orchids (the other is *Piperia elegans*) that grow in the non-native eucalyptus groves dotting California. In the Santa Rosa Mountains of Riverside County, the spotted coral-root grows under willows. It can grow in seemingly impossible conditions, sometimes pushing up through the gravel-and-tar mixture lining roads, and blooming just a few centimeters off the pavement.

Blooming season: *Corallorhiza maculata* has one of the longest blooming seasons of California's orchids, and is also one of the first to bloom. Those along the coast open as early as late February. Inland, the onset of blooming is delayed, and in the mountains flowering lasts into early August. Flower spikes emerge over a long period, and it is possible to see ripening seed capsules, stems in full bloom, and emerging spikes in the same area at the same time, a pattern that results in several months of blooms in many locations.

The spotted coral-root blooms with many of our other orchids, often side by side with *Corallorhiza striata* and in the same area as *C. mertensiana*. It blooms intermixed with *Cephalanthera austiniae, Cypripedium fasciculatum,* and *C. montanum. Calypso bulbosa, Goodyera oblongifolia,* and many of the *Piperia* grow nearby, but bloom at different times.

Conservation: The spotted coral-root is one of the most abundant orchids in the state, and is adaptable to diverse habitats; consequently, it appears to be safe from threats. Although habitat is being lost, primarily from logging operations, major populations are protected within State Parks and National Parks, and in wilderness areas.

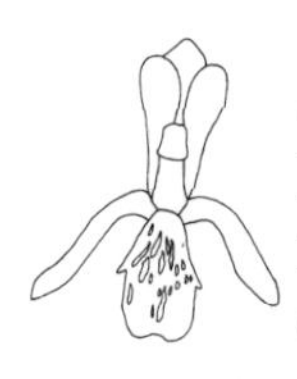

Corallorhiza mertensiana Bongard

Mémoires Academie Impérial Science St. Pétersbourg VI 2: 165. 1832.

Etymology: This plant was named after Franz Carl Mertens, a German botanist.

Synonymy:
Neottia mertensiana (Bongard) Kuntze, Reviso Generum Plantarum 2: 674. 1891.
Corallorhiza vancouveriana Finet, Bulletin Société Botanique de France 56: 100. 1909.
Corallorhiza purpurea L. O. Williams, Bulletin Torrey Botanical Club 59: 427. 1932.
Corallorhiza maculata Rafinesque ssp. *mertensiana* (Bongard) Calder and Taylor, Flora of the Queen Charlotte Islands: 288. 1968.

Common names: western coral-root, Mertens' coral-root.

Plate 5

Description: Larger plants of *Corallorhiza mertensiana* (mur - ten - zee - ah′ - na) exceed 50 cm in height with upwards of 30 flowers, but plants as small as 9 cm tall can bloom. The flowers are typically 1.5 cm in vertical extent. Bright-pink or purple coloring dominates the entire plant, including the leafless stems and the bracts. The oblanceolate sepals and petals have various amounts of brown or yellow at the tips, sometimes along their full length. The elliptic-ovate lip, typically about 0.4 × 0.8 cm, is the most colorful part of the flower, sometimes solid purple, sometimes purple mixed with white. The three-lobed lip turns down sharply near its base, and for most of its length is held parallel to the inflorescence axis. The lateral lobes of the lip vary from plant to plant, even in the same area. On some they resemble smaller versions of the lateral lobes of *C. maculata,* with prominent forward projections; on others, the lobes are barely noticeable. The center lobe of the lip supports two raised ridges. The column, long, slender, and slightly curved, is longer than those on the other coral-root species in California. The column and upper petals are aligned with the outline of the dorsal sepal, giving the appearance of one part instead of four. The lower sepals are sometimes held out to the side, but as often are folded back along the ovary. The bases of the lower sepals extend backwards along the ovary to form a short mentum, one that is more pronounced than that in *C. maculata.* The capsules of *C. mertensiana* are ellipsoid and pendant. Calder and Taylor (1968) consider *C. mertensiana* only a subspecies of *C. maculata,* but most botanists believe the plants are sufficiently distinct to justify maintaining specific status for each.

Yellow plants with a pure white lip, much like the color pattern of *C. maculata* var. *flavida,* have been found in portions of *C. mertensiana*'s range (see Fries, 1970; Luer, 1975; and Petrie, 1981). Notes on an herbarium collection from Mendocino County indicate that the plant was yellow, so perhaps yellow-flowered *C. mertensiana* also grow in California.

Distribution: *Corallorhiza mertensiana* grows along California's west coast into Canada, where it spreads east. A finger of population extends down into Wyoming, and the plant is easily found in Yellowstone National Park. This distribution leads to the common name western coral-root. In California, *C. mertensiana* occurs in only nine counties. In the northwestern corner, it extends along the coast from Del Norte County down to Mendocino County, and inland into Siskiyou and Trinity Counties. There is also a population in the Sierra Nevada in Butte, Plumas, and El Dorado Counties. Because of habitat similarities, the western coral-root may also be expected to occur in the mountainous Sierra, Nevada, and Placer Counties, but so far has not been documented there.

Habitat: *Corallorhiza mertensiana* grows in mixed coniferous forest along the coast and in the mountains, from near sea level to nearly 2200 meters. It seems to thrive with other plants, growing among irises and ferns, in mosses, and totally concealed under shrubs. Its most common habitat, particularly away from the coast, is the litter of conifers, often in dense shade. Like C. *maculata,* it sometimes forms large colonies.

Blooming season: The blooming season for the western coral-root begins in early May, when the plants in the coastal counties start to open. First bloom in the inland counties comes later, extending the species' season into early August. The first three weeks in June, when blooming plants are easily spotted from a considerable distance, are the best part of the season. The western coral-root is sympatric with C. *maculata,* and also blooms with *Calypso bulbosa* and *Cephalanthera austiniae. Listera caurina, Goodyera oblongifolia,* and *Piperia unalascensis* also bloom nearby.

Conservation: The primary threat to C. *mertensiana* is loss of habitat due to logging, but because large populations exist in protected habitats such as State Parks and designated Wilderness Areas, it appears to be safe at this time.

Notes and comments: The western coral-root was my first wild orchid. My wife, Jan, and I came across it in Olympic National Park in June 1972. Until then I was not aware there were wild orchids in the United States to rival the commercial orchids in both beauty and intrigue. That chance discovery prompted an enduring pursuit, and in each ensuing year at least part of every vacation and many weekends have been devoted to looking for our wild orchids. Simply by growing along a trail, C. *mertensiana* started the process that led to this book.

Corallorhiza striata Lindley

Genera and Species of Orchidaceous Plants: 534. 1840.

Etymology: The epithet *striata* is from the Latin word for striped.

Synonymy:
Corallorhiza macraei A. Gray, Manual of Botany, ed. 2: 453. 1856.
Corallorhiza bigelovii Watson, Proceedings American Academy of Arts and Sciences 17: 275. 1877.
Neottia striata (Lindley) Kuntze, Reviso Generum Plantarum 2: 674. 1891.
Corallorhiza vreelandii Rydberg, Bulletin Torrey Botanical Club 28: 271. 1903.
Corallorhiza ochroleuca Rydberg, Bulletin Torrey Botanical Club 31: 402. 1904.

Common name: striped coral-root.

Plate 6

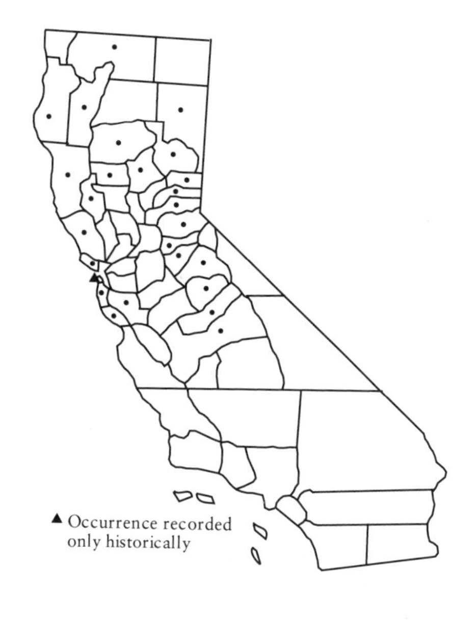

Description: Plants of *Corallorhiza striata* (stry - ah′ - ta) exceed 40 cm in height, with upwards of 45 profusely striped flowers borne on the leafless stems. Each flower is about 1.6 cm across. The sepals are oblong-elliptic, and the petals are smaller and more linear. The sepals and petals, which have three to five brownish to red veins, remain slightly cupped forward. The lip is entire, concave, and ovate-elliptic, with five stripes and a bilobed callus near the middle of the base. The background color of the flowers is brown to reddish-brown. The column is slender and slightly arched. The striped coral-root is the only member of the genus in California that lacks a mentum. The capsules are ellipsoid and pendant.

Multiple color forms of *C. striata* have been given varietal names. *Cor-*

allorhiza striata Lindley var. *vreelandii* (Rydberg) Williams has a light-tan to yellowish base color and dull-brown stripes. Its color is generally much less intense than that of the more common flowers described in the preceding paragraph, and it is sometimes described as smaller, as well, although there is considerable overlap in flower size. Rydberg (1903) had originally named this variety *C. vreelandii* in honor of a field companion, and some authorities have granted specific status to the pale color form, calling it *C. bigelovii* Watson (see Abrams, 1940). Abrams' recognition, however, was qualified, because in comparing the form to *C. striata,* he said it was "doubtfully distinct from the next."

At least two other color varieties of *C. striata* have been named. *Corallorhiza striata* f. *fulva* Fernald has sheaths and perianth of yellow or orange-brown. This description matches the yellow flowers with red stripes, and the gold flowers with red stripes, that grow in scattered locations within California. *Corallorhiza striata* f. *fulva* is the most common form of *C. striata* in New Mexico (see Todsen and Todsen, 1971). Pure-yellow plants and flowers have been called either *C. striata* var. *flavida* Todsen and Todsen or *C. ochroleuca* Rydberg. So far, pure-yellow plants of *C. striata* have not been documented in California, but Correll (1978) reported a yellow form of *C. striata* on the west coast, so they may also occur here. As with *C. maculata,* there is almost continuous variation in shading between the named forms of *C. striata,* and varietal distinctions are therefore questionable.

Distribution: *Corallorhiza striata* is widely distributed, ranging from California into Canada and across the northern states and provinces, with a disjunct location in Mexico. Within California it occurs in 25 counties. Along the coast its southern limit is Santa Cruz County, and in the Sierras it extends as far south as Tulare County. Because suitable habitat exists in Del Norte, Shasta, Modoc, and Napa Counties, the striped coral-root most likely also grows there. The species was once reported from the western part of the city of San Francisco (see Howell et al. 1958), but there are neither recent references nor herbarium specimens to indicate that it still occurs there. Most likely, the plants that Howell et al. mentioned fell victim to urbanization. There is a single herbarium specimen of *C. striata* purported to have been collected in 1875 from San Diego County. In the absence of supporting data, the possibility that the striped coral-root occurs that far south in California has been discounted, but merits additional field studies.

Habitat: The striped coral-root grows predominantly on the dry, open floor of both evergreen and mixed deciduous forests between 75 and 2200 meters. In the coastal mountains it grows in oak woodlands, either in the open or in dense patches of ferns and poison oak. Like C. *maculata,* it sometimes grows in moist environments near streams.

Blooming season: The striped coral-root is the first wild orchid to bloom in California, opening just a few days ahead of C. *maculata.* It has a blooming season about as long as that of C. *maculata,* but the latter peaks slightly earlier. In the Santa Cruz Mountains it flowers first in late February. Blooming occurs primarily in May and June in the Sierras, with a few stragglers lasting until the end of July. *Corallorhiza striata* can put on spectacular flowering displays; large colonies sometimes form, and all of the plants open at the same time, making it possible to see hundreds of blooming plants from a single spot. Through much of the northern portions of its range, C. *striata* is sympatric with C. *maculata.* Other orchids that often bloom nearby include *Epipactis helleborine, Goodyera oblongifolia, Calypso bulbosa,* and *Cephalanthera austiniae.*

Conservation: Because of its wide distribution and the presence of large populations inside the boundaries of State Parks and National Parks, the striped coral-root appears well protected from threats at this time.

Corallorhiza trifida Chatelain var. *verna* (Nuttall) Fernald

Rhodora 48: 196. 1946.

Etymology: The epithet *trifida* derives from the Latin word for "split into three," in reference to the three-lobed lip.

Synonymy:
Corallorhiza verna Nuttall, Journal Academy of Natural Sciences, Philadelphia 3: 136. 1823.
Corallorhiza innata var. *virescens* Farr, Transactions Botanical Society of Pennsylvania 2: 425. 1904.
Corallorhiza corallorhiza ssp. *coloradensis* Cockerell, Torreya 16: 231. 1916.
Corallorhiza wyomingensis Hellmayr and Hellmayr, Rhodora 33: 133. 1931.

Common names: early coral-root, pale coral-root, northern coral-root.

Plate 7

Description: *Corallorhiza trifida* (try - fid′ - ah) is the smallest, and has the least flower production, of the four coral-root species in California. Mature plants are typically only 14 cm tall and carry about 10 flowers. The leafless raceme, the sepals, and the petals are a uniform chartreuse green. The pure-white lip is three-lobed, as in *C. maculata,* but lacks the slight spreading at the tip, and the lateral lobes of the lip are not as long. The linear-lanceolate sepals are proportionately longer than those in *C. maculata,* extending below the lip in a slight curve. The petals are oblanceolate, and remain partially closed about the column. The capsules are ellipsoid and pendant. The green coloration of *C. trifida* var. *verna* demonstrates that chlorophyll, absent or meager in the other coral-roots, is present in relatively large amounts. The typical color form found in the northeastern

United States is often slightly tinged with brown on the sepals and petals, and may have purple spots on the lip. The light-green flowers with pure-white lips, such as we have in California, were originally segregated as *C. verna* Nuttall, but were subsequently reduced to variety status. Wilken and Jennings (in Hickman, ed., 1993) do not recognize var. *verna*. Because color variation in *C. trifida* is much more limited than in either *C. maculata* or *C. striata*, and because the color variants are more geographically isolated, the use of var. *verna* has merit and should be continued.

There are no reports about the pollination mechanism of *C. trifida* in the United States, but European plants are self-pollinated when the pollinia fall onto the stigmatic surface (see Davies et al., 1988).

Distribution: Though it is the only member of the genus occurring in Europe or Asia, the early coral-root is well known in many parts of the world. It is circumpolar in distribution and quite common, occurring in large colonies in several parts of its range. *Corallorhiza trifida* var. *verna*, the variety that occurs in California, is found in the southern limits of the species' distribution. Although found fairly frequently in the eastern and northern parts of the United States, the early coral-root is rare in the western states. It occurs in only one location in Nevada (see Sorrie, 1978), and is rare in the Blue and Wallowa Mountains of Oregon (see Eastman, 1990). *Corallorhiza trifida* is also one of the rarest orchids in California, its known distribution limited to a small portion of Plumas County.

Habitat: The Plumas County habitat of *C. trifida* is a meadow, at about 1500 meters elevation, that is in transition to forest. The encroaching forest is a mixture of conifers and deciduous trees. The coral-roots grow in grasses and mosses near the edges of the meadow, in the protection of pines, firs, and aspens. The known population consists of only five scattered groups, although because of its small size and its propensity to emerge in grasses, *C. trifida* is probably often overlooked, and there may be more plants bordering the meadow area.

Blooming season: Though known as the early coral-root on the East Coast because it begins blooming in April, *C. trifida* is the last coral-root to bloom in California. The plants bloom in late June and early July. By then, *C. striata* and *C. mertensiana* are nearly finished blooming, and *C. maculata* has been blooming for months. The relatively short documented blooming season for the early coral-root may be due simply to lack of exposure. As we learn more about the plant in California, and perhaps discover it in more locations, the limits of the known blooming season will most likely increase. Because var. *verna* blooms in a relatively damp area, it is not surprising that the other orchids that bloom nearby—*Platanthera dilatata* var. *leucostachys, P. sparsiflora,* and *Listera convallarioides*—are also those that grow in damp habitats.

Conservation: With its known distribution in California restricted to a small area of only one county, *C. trifida* is at risk of extirpation due to loss of habitat from logging, fire, or natural phenomena such as flooding or landslides. Fortunately, the U.S. Forest Service has shown some interest in protecting its habitat, and, barring a natural disaster, there is reason to hope for the continued existence of *C. trifida* in California.

Notes and comments: The possibility that *C. trifida* might occur in California was first brought to my attention by Bill Netherby, a fellow wild-orchid fancier. Bill related finding it several years ago in Plumas County. He tried to relocate it the year following his discovery, but could not. Using Bill's data I searched several areas of Plumas County in 1988, but was unable to find the plant. My next encounter with *C. trifida* came in late 1989 while I was doing some research at the herbarium of the California Academy of Science in San Francisco. The Academy's collection includes a specimen and photographs of the early coral-root submitted in 1977 by Bruce Sorrie and Dennis Beall. They had found *C. trifida* in Plumas County in early July. Sorrie (1978) subsequently published an account of their discovery. The evidence of the discovery by Sorrie and Beall renewed my determination to find this orchid in California.

In July 1990 I enlisted the aid of Leon Glicenstein, and we spent two days searching the borders of meadows in the area where Sorrie and Beall had made their collection before finding two tiny plants of *C. trifida* poking through some moss. The orchids were at the bottom of a moist channel, close to its origin. The plants were growing with corn lilies (*Veratrum* species), lupine, and grasses, under an open canopy of firs. Only the two plants were clearly visible, but by searching on hands and knees through the grasses we were able to locate six more. Subsequent discoveries increased the known colonies around the meadow to the five previously mentioned.

5. *Cypripedium* Linnaeus

Species Plantarum 2: 951. 1753.

Etymology: *Cypripedium* is from the Greek words for Aphrodite (the goddess of love) and foot, and would closely translate as Aphrodite's foot, hence lady's slipper.

The genus *Cypripedium* (sip - ri - pee′ - dee - um) comprises more than 30 species distributed broadly in the Northern Hemisphere. Eleven species grow in the United States, three of these in California. The plants are perhaps more readily recognized by the common names of lady's slipper or moccasin flower. Both names derive from the distinctive floral pouch that in some species clearly resembles a slipper in profile.

There are four genera in the subfamily Cypripedioideae, all of which share the common name slipper orchids because of the shape of their lip. Atwood (1984) gives a detailed description of the differences separating the genera. *Cypripedium* and *Selenipedium* have thin plicate leaves along the stem. *Phragmipedium* and *Paphiopedilum* have thick, conduplicate, basal leaves. *Cypripedium* and *Selenipedium* are differentiated on the basis of their ovaries: *Cypripedium* has a single-cell ovary; *Selenipedium,* a three-cell ovary. Only *Cypripedium* occurs in the United States.

Most lady's slippers have a structural modification in addition to the pouch: they appear to have only two sepals instead of the usual three, because the two lower sepals have united into a single unit called the *synsepal* (often a notch at the tip of the synsepal hints at its origin). The lady's slippers have two fertile stamens, one on each side of the column, and a staminode, or sterile stamen. The pouch, a modified lip, plays a major part in pollination by temporarily trapping visiting insects and channeling them such that they must crawl out past the stigma and one of the anthers. In the struggle to exit, the insect comes in contact with a pollen mass, and carries it away. The pollen is deposited in the next flower visited, effecting cross-pollination. Reproduction in *Cypripedium* thus involves pollination by deceit, because there are no food rewards in the flowers, only an enticing scent.

Three lady's slippers are well established in California: *C. californicum, C. fasciculatum,* and *C. montanum.* One piece of evidence indicates that a fourth may be, or may have been, in the state. An undated specimen from Sierra County housed at the Jepson Herbarium in Berkeley was identified in 1988 by Charles Sheviak of the New York State Museum as *Cypripedium calceolus* var. *parviflorum,* and later modified to *Cypripedium parviflorum Salisb.* var. *makasin* (Farwell) Sheviak (pers. communication). That single specimen sheet is the only record indicating that a yellow lady's slipper may once have occurred in California. An extensive search of the collection location noted on the specimen sheet failed to turn up any yellow lady's slippers, and it seems unlikely, though not impossible, that such a showy flower could escape notice except on that single occasion. Perhaps the unknown collector made an error when recording the location on the collection sheet (errors of this type are not unusual), or perhaps *C. parviflorum* var. *makasin* has been extirpated from California. The happiest outcome would be to discover that *C. parviflorum* var. *makasin* does hide extremely well and is still in the Sierra Nevada, but because the evidence is so scant, *C. parviflorum* var. *makasin* is not today included among the Orchidaceae of California. *Cypripedium parviflorum* var. *makasin,* a plant widely distributed in the northern United States and Canada, was only recently described (Sheviak, 1993). Previously var. *makasin* had been included in the *C. calceolus* complex of yellow lady's slippers.

None of the three extant lady's slippers are on the State of California list of endangered, threatened, or rare plants (State of California, 1987), but all three are on the Watch List (Smith and York, 1984) maintained by the California Native Plant Society. Inclusion on the Watch List means the plants are potentially threatened, but currently exist in sufficient numbers and with wide enough distribution to be safe from extinction. Throughout their range, however, the lady's slippers are threatened by logging and collecting.

Key to the California Species of *Cypripedium*

1. Leaves two, midway on plant; flowers clustered — *C. fasciculatum*
1. Leaves more than two, scattered along the stem; flowers on a raceme:
   2. Sepals and petals about the same length as the lip, yellow to yellow-green; wet places — *C. californicum*
   2. Sepals and petals much longer than the lip, usually brown to tan; dry places or near the edge of wet places, but not in wet places — *C. montanum*

Cypripedium californicum A. Gray

Proceedings American Academy of Arts and Sciences 7: 389. 1868.

Etymology: *Cypripedium californicum* was named for the state in which it was discovered.

Synonymy: none.

Common name: California lady's slipper.

Plate 8

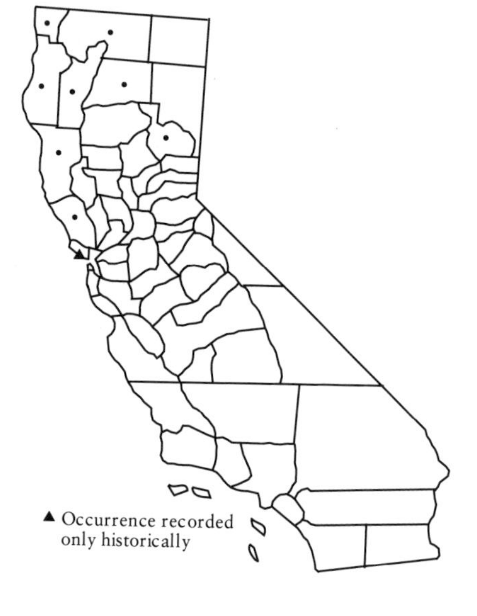

Description: *Cypripedium californicum* (kal - i - for′ - ni - kum) is the tallest of the three cypripediums in California, although its size varies considerably. Blooming plants approach 130 cm in height but plants under 25 cm are common, and in both Mendocino and Plumas Counties some plants consistently bloom at as little as 8 cm. Typical mature plants will have seven or eight alternating thin, plicate leaves approximately 10 × 14 cm and carry up to 14 flowers, each measuring 2.5 × 3.8 cm. Flower color is remarkably constant throughout the range. The white pouch is about 1.2 × 1.8 cm, and sometimes has faint red to purplish veining on the inside, and some yellow on the outside bottom. The inside rear of the pouch is lined with fine white hairs. The sepal, synsepal, and petals are yellow to greenish yellow, and covered with fine hairs. The elliptic dorsal sepal is about 1.3 × 0.8 cm, and the synsepal is 1.0 × 1.7 cm. The petals are lanceolate-oblique, and about 0.5 × 2.0 cm. The staminode is pure white except for a vertical green stripe at its center. Each flower node supports a leaflike bract in addition to the flower. When a flower first opens, the petals are reflexed back somewhat, but straighten out as the flower ages. The flowers have a faint, slightly sweet aroma. The capsules are ellipsoid. In addition to propagating by seed, *C. californicum* propagates asexually, by sending up multiple new growths from its fibrous rhizome, which usually has multiple fibrous roots.

Distribution: *Cypripedium californicum*, limited to northern California and portions of southern Oregon, has the smallest range of any of the lady's slippers in the state. Within California it occurs in nine counties, reaching as far south as Sonoma County along the coast, and into Plumas County in the Sierras. An unconfirmed Forest Service record of *C. californicum* in Butte County suggests that the lady's slippers may occur there as well. Unfortunately, *C. californicum* seems to be withdrawing from some of its former range. It once occurred in Marin County (see Howell, 1970), but repeated attempts to locate it there have been unsuccessful. The species was described on the basis of material collected from Red Mountain in Mendocino County.

Habitat: Of California's three lady's slippers, *Cypripedium californicum* favors the wettest conditions, requiring a constant supply of water at its roots. The plants are typically found on the banks of streams or in hillside seeps between 60 and 2130 meters elevation. Another favored spot is boglike areas close to the water's edge. They grow almost exclusively in and among serpentine rocks or in serpentine-based soils, often in full sun. Azaleas and dogwoods are frequent companions, and it is common to find the orchid foliage concealed under azaleas, with only the flowers reaching above the branches. *Cypripedium californicum* typically occurs in larger colonies than do our other two lady's slippers. Blooming clumps often exceed a hundred stems, with hundreds of additional stems nearby. The California lady's slipper frequently grows in close association with the unusual *Darlingtonia californica,* an insectivorous plant also endemic to serpentine substrates in northern California and southern Oregon. Its pitcher-shaped leaves trap and devour insects to compensate for the nutrient deficiencies in the serpentine soils. The massive colonies of

D. californica are easier to spot than are the lady's slippers, and are a good indication that *C. californicum* may be nearby.

Blooming season: *Cypripedium californicum* is the last of our lady's slippers to open. The blooming season starts in early April in the coastal mountains. In the Cascade Range, blooming is going strong by early June, and flowering lasts to the end of July in the Sierras and at high elevations in the Cascades. Several other orchids often bloom with *C. californicum.* The tiny, green-flowered *Platanthera sparsiflora* often forms large colonies on rivulets and seeps, along with the lady's slippers. Along some streams, *Epipactis gigantea* and *Platanthera dilatata* bloom intermixed with the lady's slippers. *Corallorhiza maculata, C. striata, Piperia transversa, P. unalascensis,* and *Goodyera oblongifolia* grow in drier areas near the lady's slippers.

Conservation: Unfortunately, owing to their beauty, the lady's slippers are often prey for collectors. The unmistakable signs of digging are frequently found in the more easily accessible colonies. Logging is not as great a threat to *C. californicum* as it is to the other two lady's slippers in California. The hillside seeps they favor usually support very little harvestable timber, and recent Forest Service practice has been to protect riparian areas, another favorite habitat. *Cypripedium californicum* is considered rare in Oregon (see Eastman, 1990).

Notes and comments: In northern California the beginning of the blooming period of *C. californicum* may coincide with the latter part of the rainy season. Sometimes, after driving for nearly two days, one finds that it is raining during the only time available to hunt for the orchids. Rather than turn back without searching, just pull on rain gear, and set out to find the lady's slippers. It is well worth it to spend a few extra dollars investing in really good rain gear in order to make these early-season orchid hunts successful and bearable.

Cypripedium fasciculatum Kellogg ex S. Watson

Proceedings American Academy of Arts and Sciences 17: 380. 1882.

Etymology: The epithet *fasciculatum,* from Latin, means "gathered into a bundle," here in reference to the clustering of the flowers.

Synonymy:
Cypripedium pusillum Rolfe, Kew Bulletin: 211. 1892.
Cypripedium knightae A. Nelson, Botanical Gazette 42: 48. 1906.

Common names: clustered lady's slipper, small brown lady's slipper.

Plates 9 and 10

Description: *Cypripedium fasciculatum* (fa - sik - yoo - lah′ - tum), by far the shortest of California's lady's slippers, is one of the "belly orchids." To really enjoy it, you have to plop down on the ground on your belly because the plants are under 18 cm from the base to the tops of the flowers. *C. fasciculatum* has a small rhizome with fibrous roots. The two opposed leaves, halfway up the hairy stem, are plicate, elliptic, and fuzzy on the underside. Most plants have a total leaf span of about 30 cm, but plants just 3.2 cm tall with a leaf span of 9.5 cm are capable of blooming. There is usually a single miniature bract between the leaves and the flowers. On the rare three-leaved plants, the third leaf is between the primary pair and the bract. The flowers are small, only about 4.5 cm from tip to tip, and sometimes much smaller, especially on multi-flowered plants. The dorsal sepal and the synsepal are lanceolate-ovate. The petals are ovate-acuminate. The pouch measures about 1.0 × 1.3 cm, with an opening about 0.5 × 0.6 cm. Tiny hairs line the inside rear of the pouch. The pouch, sepal, synsepal, and petals are usually all the same color, but sometimes the pouch is lighter than the rest of the flower. The staminode is usually a light green, but on dark flowers has a slight brownish tinge. Usually, the flowers have brown markings on a green or golden background, although considerable variation in color exists. Sometimes the brown dominates, and the flowers appear dark brown to nearly red. Other plants have very pale, almost blond flowers, and a few plants have pure-green flowers. The flowers, typically borne in clusters of six or seven, but sometimes as many as ten, form a clump at the end of the stem. Multiple flowers seemingly overburden the stem, causing it to droop, one of the reasons you have to get down low to see them. As the obovoid-ellipsoid capsules mature, the stem straightens out, presenting the capsules in a semi-erect position.

The clustered lady's slipper propagates by both asexual and sexual means. It is common to find tightly grouped plants bearing identical flowers that appear to be clonal propagation. One such group contained 33 blooming plants in a space of only 25 × 43 cm. Seedlings spread widely around mature plants are evidence that seed propagation is also occurring.

Distribution: *Cypripedium fasciculatum* is perhaps the most interesting of the California lady's slippers because of its distribution. A majority of references imply a continuous occurrence throughout its range. Typically, it is reported from the northern half of California into Oregon, Washington, Montana, Utah, Idaho, Colorado, and Wyoming. Within its range, however, it occurs, not in a continuous distribution, but in disjunct populations (see Brownell and Catling, 1987). The plants in the eastern edge of the range were once thought to differ from those on the west coast, and were named *C. knightae* Nelson, but Brownell and Catling maintain there are no substantive differences in plants from the two regions. Three major pockets of plants, covering 13 counties, occur in California. These pockets are in the northwestern counties of Del Norte, Trinity, Tehama, Humboldt, and Siskiyou; the Sierran counties of Butte, Plumas, Yuba, Sierra, and Nevada; and the Santa Cruz Mountains counties of Santa Cruz, San Mateo, and Santa Clara.

Habitat: The clustered lady's slipper grows mainly in fir forest between 170 and 1980 meters elevation. Most often the plants grow on or near a stream bank, often on fairly steep slopes, or just above the drainage on the forest floor, either in the open or under dogwoods. The plants growing under dogwoods are usually totally obscured late in the season after the trees leaf out. Some major colonies are set back from streams more than 100 meters. Plants nearer streams grow in competition with other riparian vegetation, such as *Clintonia uniflora,* or under complete shrub cover, where they can be nearly impossible to find. Occasionally, plants grow on moss-covered rocks at the edges of streams, where they root in cracks in the rocks. In low-rainfall years such as we experienced during the extended drought that began in the late 1980s, the streams supporting the lady's slippers may dry completely before flowering commences, but the quality of blooms remains high. Most often, the clustered lady's slippers grow as scattered individuals or small groups, although large clumps do occur. A 2-square-meter area in Plumas County contained over 150 blooming plants. One of the most unusual habitats of *C. fasciculatum* is steep roadcuts, where the plants are clearly visible to passing traffic. Possibly the slope and openness of the roadcuts approximate the conditions of the favored streamside habitat.

Blooming season: *Cypripedium fasciculatum* is the first of our lady's slippers to bloom, and has the longest blooming season. Flowering starts in mid-March in the Santa Cruz Mountains (if the species is still extant there). Farther north, blooming begins in early May and lasts into June. At higher elevations in the Sierras, peak blooming occurs in mid to late June, with flowers lasting into July. In most years the season ends in early July, but the herbarium record indicates that blooming sometimes extends into late July. Many other orchids, including *C. montanum,* bloom near *C. fasciculatum. Cephalanthera austiniae, Corallorhiza maculata,* and *C. striata* often bloom at the same time as does *Cypripedium fasciculatum,* and within a few feet. *Calypso bulbosa* also blooms nearby early in the season in some locations. Piperias and *Goodyera oblongifolia* will be showing spikes, but will not bloom until long after the flowers on the lady's slippers have faded.

Conservation: The perennial nemeses of many of our native orchids, collecting and habitat destruction through logging, also threaten *C. fasciculatum,* but it is encouraging to see Forest Service recognition of this orchid. One timber sale in Plumas County was reconfigured by the Forest Service after logging had started when it was discovered that the sale area included a large colony of *C. fasciculatum.* This conservation measure saved over 1000 plants.

The clustered lady's slipper is considered rare in Oregon (see Eastman, 1990), and is rare, or perhaps no longer extant, in parts of its range in California. There are multiple herbarium collections of the plant from the Santa Cruz Mountains counties of San Mateo, Santa Clara, and Santa Cruz, but the last known siting was in 1967. Multiple searches over the last several years by members of the local chapter of the California Native Plant Society have not turned up any evidence of the clustered lady's slipper, and it seems that some reduction in range is occurring, owing to loss of habitat.

Notes and comments: During the early part of the blooming season, the weather in the mountains can be quite unpredictable, with rain likely and snow flurries possible. During a mid-May trip to see *C. fasciculatum* in Nevada County, a surprise snowstorm blanketed my campsite with about 10 cm of snow. The orchids were at a slightly lower elevation where the snow did not stick, but it fell all morning. The flowers were unaffected by the weather, and although we do not normally think of seeing orchids in the snow, with these hardy plants it can happen.

Cypripedium montanum Douglas ex Lindley

Genera and Species of Orchidaceous Plants: 528. 1840.

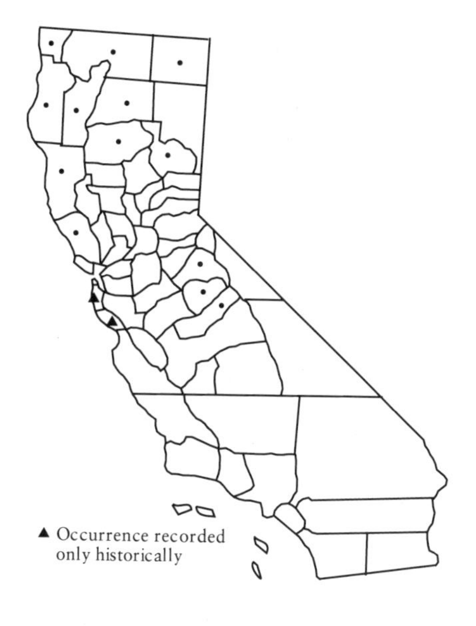

Etymology: The epithet *montanum* is from a Latin word meaning mountain, the typical habitat of this species.

Synonymy:
Cypripedium occidentale Watson, Proceedings American Academy of Arts and Sciences 11: 147. 1876.

Common name: mountain lady's slipper.

Plates 11 and 12

Description: *Cypripedium montanum* (mon - tah′ - num) blooms on plants ranging between 25 and 66 cm tall. Mature plants have five or six alternating, plicate, ovate-lanceolate leaves distributed along the stem. The slightly fuzzy leaves can be over 10 cm wide and 18 cm long. The plants sometimes bear three flowers, but most often have one or two. The largest flowers have a natural spread of approximately 10 × 10 cm, measured from petal tip to petal tip, making this the largest orchid flower in California. The spread would be even larger if the curled and drooping sepals and petals were straightened out. The pouch is white, veined with purple, occasionally with a purple rim around the opening. The veining is most readily observed by looking into the pouch, but on some plants the veining also shows on the outside. The inside rear of the pouch is lined with fine hairs that help provide an escape path for the pollinators that become trapped inside. The typical pouch measures about 1.6 × 3.0 cm, with an opening about 1.1 × 1.3 cm. The undulate dorsal sepal and the synsepal are ovate-lanceolate, and each is about 1.5 × 6.0 cm. Occasionally, plants will have three fully developed sepals instead of a sepal and a synsepal. The twisted petals are linear-lanceolate and about 0.5 × 5.0 cm. The color of the sepals and petals varies from an intense brown to a light tan. The petals are slightly hairy near the pouch, and the backs of both sepals and petals are covered with fine hairs. The staminode is a bright yellow with red spots, providing a brilliant contrast to the brown sepals and petals and the white pouch. The flowers have a pleasant fragrance, slightly reminiscent of anise. The ellipsoid capsules are held erect to semi-erect. The plants spread both by seed and by multiple leads from short rhizomes and fibrous roots.

Distribution: *Cypripedium montanum* has a wider range, both inside and outside the state, than do the other lady's slippers that occur in California. In addition to California, it grows in Wyoming, Montana, Idaho, Oregon, Washington, Alaska, British Columbia, and Alberta. Within California, *C. montanum* occurs in 15 counties, reaching as far south as Santa Cruz County along the coast, and down into Madera County in the Sierra Nevada, although it is not continuous within this range. For example, along the coast it apparently skips Marin and San Francisco Counties, and in the Sierras jumps from Plumas County down to Tuolumne County. There are many herbarium records of the species from the counties of Tuolumne, Mariposa, and Madera, so it appears strange that it has not been collected in some of the central Sierran counties that contain seemingly perfect habitat. Within its known area, populations range from extremely rare, in the Santa Cruz Mountains, to locally plentiful, in pockets of the Sierra Nevada.

Habitat: *Cypripedium montanum* has adapted to multiple habitats, growing in both moist and dry conditions at elevations between 180 and 2130 meters, although it is rare above about 1500 meters. The typical moist condition favored by the plant is near a stream, sometimes on the bank of a small stream or on the edge of a hillside seep. In these moist conditions it grows under or near the creek dogwood, *Cornus stolonifera,* and the Western azalea, *Rhododendron occidentale. Clintonia uniflora,* a small, white-flowered member of the lily family, often forms a continuous ground cover around the mountain lady's slippers. Less often, *C. montanum* grows in or adjacent to marshy areas. In these wetter conditions it grows with ferns, lilies, and corn lilies under the protection of alders, firs, and cedars. *Cypripedium montanum* also grows in relatively dry conditions on hillsides in mixed and coniferous forest, under oaks, firs, and madrones. In the more southerly parts of its range, the mountain lady's slipper will grow in poison oak. Occasionally, plants are found in nearly full sun, though usually they are in at least partial shade. Often, clumps of multiple blooming stems develop. An especially dense group in Plumas County consisted of over 80 stems, with 70 of them in bloom, all in an area of under 1 square meter. Like *C. fasciculatum, C. montanum* sometimes shows up on roadcuts, where the white pouch makes it easy to spot.

Blooming season: The blooming period of *C. montanum* varies with elevation. The herbarium data show blooming beginning in late March in the lower elevations, such as in the Santa Cruz Mountains. Blooming begins in early May in the Sierras, continuing into the first few weeks of June. Some years, blooming may last through early July at the upper limit of the elevation range. Several other orchids bloom at the same time and in the same locations as does *C. montanum: Cephalanthera austiniae, Platanthera dilatata, P. sparsiflora, Corallorhiza maculata,* and *C. striata.* Plants of *Goodyera oblongifolia, Piperia transversa,* and *P. unalascensis* not yet in bloom grow nearby. Occasionally, *C. montanum* and *C. fasciculatum* grow within reach of each other. In the wettest of its growing areas, *Listera convallarioides* blooms within a few meters of the mountain lady's slipper.

Conservation: Threats to *C. montanum* include collection and habitat destruction through logging. A particularly destructive case occurred in Siskiyou County. A pre-logging survey by the Forest Service found 260 and 300 mountain lady's slippers in a planned clear-cut area. A search of the area several years after the logging showed that only five plants had survived, on the edge of the clear-cut.

Cypripedium montanum may no longer exist in its historical range in the Santa Cruz Mountains counties of Santa Cruz and San Mateo. The most recent herbarium collection from these counties is from 1946. Unfortunately, none of the herbarium sheets from that area give precise locality data, and some of the referenced place names are no longer used. As with C. *fasciculatum* in the same region, multiple searches of the traceable collection locales have been unsuccessful in locating plants. Perhaps C. *montanum* still exists in the Santa Cruz Mountains, and is just missing from the historically documented locations, but the pressures of population growth are resulting in loss of habitat in these mountains, and in an apparent reduction in the range of C. *montanum*.

Notes and comments: One of my unfulfilled goals is to photograph a pollinator visiting C. *montanum*. I have seen and photographed visiting insects, but have not witnessed pollen being removed or deposited. For several years, though, I have kept track of successful pollination of the mountain lady's slipper at three large C. *montanum* colonies, one each in Madera, Mariposa, and Tuolumne Counties, the three sites mutually separated by about 50 km. One site was observed in four consecutive years, another one three times, the third site only twice. The success rate varied from site to site and from year to year, hitting a peak of 83% at site 1 in 1992, and a low of 17% at site 3, also in 1992. The overall success rate is 61%. Successful pollination was estimated by the simple method of counting firm green seed capsules on the plants approximately six weeks after blooming. No attempt was made to evaluate seed viability directly. The data from the study are listed in Table 4.

Table 4. Seed-capsule formation in *Cypripedium montanum*, by capules per flower, at three sites

	Site 1		Site 2		Site 3	
Year	Total flowers	Total capsules	Total flowers	Total capsules	Total flowers	Total capsules
1989	83	59	92	31	69	31
1990	85	55				
1991	89	69	109	66		
1992	103	86	116	78	46	8

6. *Epipactis* Swartz

Kongl. Svensk. Vetensk. Acad. Nya Handl. 21: 231. 1800 in part; emend. L. C. Richard, De Orchideis Europaeis Annot.: 29. 1817; Mémoires du Muséum d'Histoire Naturelle Paris 4: 51. 1818.

Etymology: According to Correll (1978), *Epipactis* is derived from a classical name used by Theophrastus for a plant used to curdle milk.

Epipactis (ep - i - pak′ - tis) is a worldwide genus of approximately 25 species. Most *Epipactis* species are native to Europe and Asia, but two species occur in the United States, and both grow in California. *Epipactis gigantea* is native and widely distributed in the western parts of the country, and is the most common orchid in California. *Epipactis helleborine* is an introduced species that appeared in California relatively recently, and is still spreading.

Epipactis is very similar to *Cephalanthera,* with flowers in both genera having three-lobed lips severely restricted in the middle. According to Davies et al. (1988), the two genera are most easily distinguished by the form of the ovary (straight in *Epipactis,* twisted in *Cephalanthera*) and the flower stem (present in *Epipactis,* lacking in *Cephalanthera*). Structural differences in the columns of the two genera, difficult to determine without a microscope, are discussed in the treatment of *Cephalanthera.*

Key to the California Species of *Epipactis*

1. Lip deeply three-lobed, the epichile elongated; wet places — *E. gigantea*
1. Lip not lobed, epichile broadly triangular; dry places — *E. helleborine*

Epipactis gigantea Douglas ex Hooker

Flora Boreali-Americana 2: 202, pl. 202. 1839.

Etymology: The epithet *gigantea* is from the Latin for gigantic, in reference to the large size of the flowers and plants.

Synonymy:
Epipactis americana Lindley, Annals Magazine Natural History I. 4: 385. 1840.
Peramium giganteum (Douglas ex Hooker) Coulter, Contributions U.S. National Herbarium 2: 424. 1894.
Serapias gigantea (Douglas ex Hooker) Eaton, Proceedings Biological Society of Washington 21: 67. 1908.
Helleborine gigantea (Douglas ex Hooker) Druce, Bulletin Torrey Botanical Club 36: 547. 1909.
Amesia gigantea (Douglas) Nelson and Macbride, Botanical Gazette 56: 472. 1913.

Common names: stream orchid, chatterbox, false lady's slipper, giant helleborine.

Plates 13 and 14

Description: *Epipactis gigantea* (jy - gan′ - tee - ah) varies greatly in size, depending on growing conditions. Blooming plants are typically 1 meter tall, although some will bloom at one-third that height, and others are as tall as 150 cm. The ovate-lanceolate, plicate leaves, 10 or more per plant, vary between 5 and 11 cm wide and up to 25 cm long, alternating on the stem. A single growth on a typical plant will have 12–18 blossoms, usually facing the same direction, although plants with more than 20 blossoms are common. Seed capsules form on the lower flowers before the upper buds open. Most flowers are between 4.0 and 4.5 cm wide, but some are as great as 5.5 cm in natural spread across the lateral sepals. The ovate-lanceolate, slightly concave sepals are dark green, slightly suffused with rose. The petals are ovate, and show more rose coloring, with darker veining. The deeply three-lobed basal half of the lip creates a pouchlike appearance, which is why *E. gigantea* is sometimes called the false lady's slipper. The lateral lobes of the hypochile are veined in red on a yellowish background, and the center lobe supports a stripe of reddish bumps. Usually, the front edges of the lateral lobes are rounded forward slightly; occasionally, the forward portion is extreme, nearly equaling the remaining portion of the lateral lobe. The lip is constricted and hinged near the middle, allowing the epichile to vibrate with the slightest breeze, which is undoubtedly the characteristic that gave rise to the common name chatterbox. The oblanceolate epichile has two fleshy calli and additional rose and some yellow coloring. Two armlike projections protrude from the upper portion of the column slightly below the light-green to yellow anther cap. The capsules are ellipsoid and pendant. The plants propagate by seed and by multiple new growths from the rhizome, which has multiple fibrous roots.

The stream orchid displays a variety of colors and some variation in shape. The extremes of color are strikingly different from one another, but there is a continuum between them. Occasionally, a flower has very intense pink coloration in the petals. The color of the epichile also varies, and in some colonies is pure yellow. Nearly white flowers occur, but infrequently. They are of the same size and shape as the normally colored ones, but are nearly totally lacking in red tones. Specifically, the hypochile has yellow veining instead of the normal red, but there is a touch of pink at its back, and some on the epichile, indicating that the flowers are not altogether anthocyanin-free. One group of *E. gigantea* in the Santa Monica Mountains has peloric flowers: all three petals have the shape of nearly perfect lips, each with full side lobes and callus. Other wild North American orchids exhibit peloria, including one other occurring in California, *Spiranthes romanzoffiana* (see Mousley, 1944).

The most unusual and perhaps the most beautiful *E. gigantea* was reported by Roger Rachie of the University of California Botanical Garden in Berkeley. The plants have deep-wine-red leaves and stems and grow in full sun on a serpentine formation in Sonoma County. The flowers enjoy some of the rich coloration of the leaves, showing a deeper rose tint than most. These may be the only red stream orchids in the state.

Epipactis gigantea apparently attracts pollinators by mimicking their food supply without giving any real reward. The stream orchid is pollinated by flies of the family Syrphidae. The flies are attracted to the flowers by the aroma, which mimics the honeydew smell given off by aphids (see Ross, 1988). Syrphid flies normally lay their eggs in masses of aphids, which become the food supply for their larvae. On the orchids, however, there are usually no aphids. The aroma of the flowers fools the fly into laying its eggs among the supposed aphids, and in the process it picks up pollinia and deposits them on the next flower visited. Though the fly does the orchid a favor by pollinating it, the orchid does not respond in kind: because there are no aphids for food, only a mimicked aroma, the hatched larvae are doomed. Nevertheless, the flies are evidently very busy, for it is common to find one or more eggs on the orchid flowers, and in many colonies virtually every flower sets fruit. *Epipactis gigantea* blooms in 39 months from seed (see Myers and Ascher, 1982).

Distribution: *Epipactis gigantea* grows from Canada to Mexico, from the Pacific to Texas, and has been reported from Wyoming (Lichvar, 1979). It is known from only 12 sites in Canada, where it is considered threatened (Brunton, 1986), and may have been extirpated from several of those locations. Fortunately, the stream orchid has a more favorable situation in California. It is easily our most common and most widely distributed orchid. It has been documented in 46 counties and on the Channel Islands, often in areas where no other orchids grow. In Imperial County, the largely desert region in the extreme southeast of the state, there is only one orchid: *E. gigantea.*

Habitat: *Epipactis gigantea* thrives in a variety of habitats, from barely above sea level, within reach of the ocean spray, to elevations as high as 2600 meters in the mountains. These habitats all have in common a constant source of water for the roots. The stream orchid grows in moist places in the deserts (including situations near springs in Death Valley), in seeps on cliffs overlooking the ocean, and in wet places in the mountains. Banks of streams are prime habitat, with the water lapping against and sometimes over the roots. Some of the colonies are located such that they are completely covered with water during the peak flows resulting from winter storms. The orchids survive these storms by rooting themselves in cracks in rocks in the streambed. Most streamside colonies grow where covering trees provide protection from the sun, but some, especially in the north, grow in full sun. The largest colonies are usually found along seeps where the constant moisture creates exactly the right environment. Most commonly, the seeps are found on the canyon sides of a drainage system or where a roadcut crosses an aquifer. Sometimes, stream orchid colonies reach immense proportions. One such colony in a natural seep in Los Angeles County had about 10,000 blooming stems.

Blooming season: The blooming season for *E. gigantea* is one of our longest, stretching from early March to the beginning of October. The earliest blooming occurs in the southern deserts and along the coast. The flowers are open in mid-April in the Santa Monica Mountains, but delay until late May or early June at the plant's highest elevations in the Sierra Nevada and remain in bloom for several more months. An herbarium specimen collected on the first of October from the Sierra Nevada still had unopened buds. Like that on all but two of our native California orchids, the above-ground growth of *E. gigantea* fades in the fall. Depending on the location and the weather, sometime in September or October the plants start turning a bright yellow, and at this time they are fairly easy to find. By November or December, the plants have turned a dark brown, and the faded leaves are all but impossible to spot. Even though no sign of life is showing above ground, the plants are very active below ground. The new shoots, usually three to five per plant, have begun to spread away from the old stem, and as early as New Year's day in Southern California, later in the north, the new shoots reach the surface. The orchid blooming companions of *E. gigantea* include *Cypripedium californicum* and *Platanthera dilatata* var. *leucostachys*.

Conservation: Because of its wide distribution and its tolerance for varied growing conditions, *E. gigantea* is safe from threats. Some human activities seem to expand its habitat. For example, roadcuts often expose new seeps, creating favorable new growing conditions, and because of its propensity to colonize roadcut seeps, *E. gigantea* is the orchid most likely to be seen by motorists. The recurring droughts in California cause fluctuations in the numbers of stream orchids. A colony of several hundred plants on a roadcut in Santa Barbara County declined steadily during the drought. By 1992, after six years of meager rain, the colony had dwindled to less than a dozen plants, none of which bloomed, and in 1993 the colony disappeared entirely even though the drought had ended.

Epipactis helleborine (Linnaeus) Crantz

Stirpium Austriarum Fasciculus, ed. 2, Fasc. 6: 467, fig. 6. 1769.

Etymology: The epithet, from Latin, means "like a hellebore," in reference to its similarity to some plants in the buttercup family.

Synonymy:
Serapias helleborine Linnaeus, Species Plantarum 2: 949. 1753.
Serapias latifolia (Linnaeus) Hudson, Flora Anglica: 393. 1762.
Epipactis latifolia (Linnaeus) Allioni, Flora Pedemontana 2: 152. 1785.
Epipactis helleborine (Linnaeus) Crantz var. *viridens* Gray, Botanical Gazette 4: 206. 1879.
Amesia latifolia Nelson and Macbride, Botanical Gazette 56: 472. 1913.

Common name: broad-leaved helleborine.

Plate 15

Description: Typical plants of *E. helleborine* (hel - uh - bore′ - ee - nee) exceed 1 meter in height, and superficially resemble *E. gigantea,* but the five to eight plicate leaves are generally flatter and wider. The fibrous roots of *E. helleborine* are shorter and more numerous than those in *E. gigantea.* The elongating flower spike droops slightly near the tip, and each flower is protected by a bract. Each spike carries up to twenty 1.8-cm star-shaped flowers, the flowers mostly green with a suffusion of pink on the petals. The sepals and petals are ovate, and slightly concave. The lip has a shiny, pouchlike depression in the center of the hypochile, which is a slightly darker pink than the petals and is usually covered with droplets of nectar. The lip is hinged near the middle, though not as freely as in

E. gigantea. The epichile is broadly triangular. Petrie (1981) described the lip as "shaped like a sauce boat with a pointed spout." The flowers give off a pleasant aroma. The capsules of *E. helleborine* are more spherical than those of *E. gigantea,* but are, like those, held in a pendant position.

A white form of *E. helleborine* called forma *monotropoides* (see Mousley, 1927) occurs in the Santa Cruz Mountains. The plants are normal in size and shape, but the leaves and stems are mostly white, with only slight amounts of pink, and not a trace of visible chlorophyll. Salmia (1986, 1989) reported that the white plants flower normally and set seed capsules, though at a lower rate than do green plants. Salmia also reported that the populations of white plants are variable from year to year, and that the white plants can apparently survive underground for one or more years before reappearing.

In the same area as the white plants grow plants with flowers and stems of a monotone bright pink, and pinkish leaves that wither by anthesis. Since Mousley's forma *monotropoides* had traces of pink in the stems, the pink-and-white plants are probably just color extremes of the same mycotrophic form. But *Epipactis helleborine* is not the only *Epipactis* with both photosynthetic and mycotrophic forms; Summerhayes (1968) reports apparent mycotrophic plants in *E. purpurata,* which occurs in Europe.

Light and MacConaill (1990, 1991) also studied the population dynamics of *E. helleborine,* and reported results similar to Salmia's. They studied 849 plants over a six-year period. Only two of the plants appeared every year, and most of the plants, 579, appeared only once in the six-year period. Fifty percent of the plants not emerging annually were still alive, and three plants appeared in 1989 after a three-year absence. Light and MacConaill also studied the underground growth of the plants. The perennating bud becomes visible one full year before emerging, and the size of the bud determines whether the plant will bloom the next year. Because the perennating buds appear so early, whether or not the plant will bloom the next year is determined even before the current year's growth blooms.

Four species of wasps pollinate *E. helleborine* at Owen Sound, Ontario (see Judd, 1971). In the process of lapping the nectar on the lip, the wasps come into contact with the stigma, depositing there whatever pollen they may have carried from another flower. Additional movement brings the wasps into contact with the rostellum, and when the wasp leaves the flower it takes with it the rostellum and the attached pollinia. The pollinators were *Polistes fuscatus, Vespula arenaria, V. consobrina,* and *V. vidua.* Other wasp pollinators, *Vespa germanica* and *V. maculata,* had been reported by Mousley (1927).

As Mousley found, wasps also pollinate *E. helleborine* in California. They are much larger than the flowers, and simply crawl from one flower to another, feeding on the nectar in the lip depression. They take many approaches to the nectar, including sideways and upside down; most often, they approach the flower aligned with the lip. As they feed, parts of their head come into contact with the pollinia. Mousley (1927) also reported that *E. helleborine* is capable of autogamy if not visited by pollinators. Fruit set is almost 100%. *Epipactis helleborine* will bloom from seed in under 18 months.

Distribution: *Epipactis helleborine* is the only non-native orchid growing wild in California. In the United States, its advent as a self-sustaining plant was first discovered in 1879 near Syracuse, New York, by members of the Ladies' Botanical Club of Syracuse (see Gray, 1879). Gray believed it to be indigenous, but it is now universally considered to have been introduced from Europe, where it is common. Once established, it spread rapidly, migrating to Canada by 1890 (see Correll, 1978). It then spread south and west to Missouri, Michigan, Indiana, and Montana by 1950, and to Illinois by 1954. The earliest herbarium records of it in California are from July 15, 1961, and they tell of blooming plants appearing spontaneously in gardens in the San Francisco Bay area. McClintock (1975), however, reports it in San Mateo County as early as 1950. It now occurs in 10 counties in California, and is still spreading.

Habitat: *Epipactis helleborine* grows in a variety of habitats, but almost always in semi-shaded conditions below 1220 meters. It grows equally well in soil that is very heavy, with seemingly little humus, or in lighter, more humus-laden soil. It can compete with other plants, even growing in dense stands of poison oak. It rapidly colonizes disturbed soils, sometimes appearing uninvited in the gardens of homes and parks. Davies et al. (1988) says that in part of its European range, *E. helleborine* grows in the open on sand dunes. Whiting and Catling (1986) also report it on sand dunes from Ontario. In California it grows only in the shade of the forest.

Blooming season: The main blooming season for the broad-leaved helleborine extends from mid-April to early September, though in the milder portions of its range, blooming may occur at almost any time. Some plants in a public park in Monterey County were blooming during late December 1992, and plants raised from seed in gardens in Ventura County regularly bloom in November and December. Plants in each area emerge over a long period, resulting in many months of bloom, and until relatively late in the season it is possible to see all stages of growth from a single spot. Other orchids blooming nearby at the same time include *E. gigantea, Corallorhiza maculata, C. striata, Piperia transversa,* and *P. candida*. In the Santa Cruz Mountains, *Calypso bulbosa* grows close to *E. helleborine,* but blooms much earlier.

Conservation: Technically, since *E. helleborine* is a non-native plant, it should not be discussed in the same context of threats as native wild orchids are. If we did, we would be forced to conclude, since it is actively spreading, that it is in no danger. In fact, *E. helleborine* should properly be classified as an invasive weed. The California Native Plant Society and several State Parks and preserves conduct programs to eradicate invasive plants. Fortunately for orchid lovers, however, *E. helleborine* has not yet been targeted for elimination. The white form of the species, which is relatively rare, according to Jorgensen (1982) and Griesbach (1979), is exceedingly rare in California, occurring in only one location, in the Santa Cruz Mountains.

7. *Goodyera* R. Brown

In Aiton, Hortus Kewensis, ed. 2, 5: 197. 1813.
Etymology: The genus was named in honor of John Goodyer, an English botanist.

The genus *Goodyera* (good - yer′ - ah) consists of at least 25 species distributed worldwide. Various authors list the number of species as anywhere from 25 to 80. *Goodyera* is closely related to *Spiranthes,* but has a creeping rhizome instead of tubers, and marked veining in the leaves. Four species occur in North America, but only *G. oblongifolia* is found in California. All North American members of the genus are called "rattlesnake plantain," and all of them produce a basal rosette of evergreen leaves marked with variable amounts of reticulation.

Goodyera is in the subtribe Physurinae. Members of this tribe of small terrestrial orchids are collectively referred to as "jewel orchids." The jewel orchids are known for their beautiful foliage, and some species are cultivated for the foliage alone, because the flowers are undistinguished. *Goodyera oblongifolia* offers the double bonus of interesting foliage and a lovely flower.

Goodyera oblongifolia Rafinesque

Herbarium Rafinesquianum: 76. 1833.

Etymology: The epithet *oblongifolia* is derived from Latin in reference to its oblong leaves.

Synonymy:
Spiranthes decipiens Hooker, Flora Boreali-Americana 2: 203. 1839.
Goodyera menziesii Lindley, Genera and Species of Orchidaceous Plants: 492. 1840.
Peramium menziesii (Lindley) Morong, Memoirs Torrey Botanical Club 5: 124. 1894.
Peramium decipiens (Hooker) Piper, Contributions U.S. National Herbarium 11: 208. 1906.
Epipactis decipiens Ames, Orchidaceae 2: 261. 1908.
Goodyera decipiens (Hooker) Hubbard in Olmstead, Coville, and Kelsey, Standard Plant Names: 328. 1923.

Common name: Menzies' rattlesnake plantain.

Plates 16 and 17

Description: Flower spikes on *Goodyera oblongifolia* (ob - long - i - foe′ - lee - ah) exceed 50 cm in height and carry over 30 small white flowers. There is a trace of greenish brown in the sepals, the tips of which curl back tightly. The dorsal sepal is triangular-lanceolate, the lateral sepals ovate. The petals are narrow at the base, but dilate above the middle. The lip is white and saccate, with a narrow, linear, recurved apex, rounded at the tip. On newly opened flowers the sepals and petals remain clasped about the lip, with only a slight spreading at the tip, and some flowers retain this tight appearance until they fade. In others, as the flowers mature, the lower sepals fold back partway and the relative positions of the column and lip change, allowing the full beauty of the flowers to be enjoyed. Even on the more open flowers, however, the dorsal sepal and petals remain connivent, forming a hood over the column and lip. The raceme is decidedly one-sided, with all the flowers facing more or less in the same direction. Six to eight oblong-elliptic leaves form a basal rosette. The individual leaves typically measure 3.5 × 8 cm, and large plants have a total leaf span of approximately 20 cm, although plants with leaf spans of just a few centimeters are capable of blooming. The leaf pattern, with its white veining, is so distinctive that the plants are easily recognized, even out of flower. The veining, or reticulation, varies considerably from plant to plant. In some plants the leaves almost lack veining, whereas in others they are very heavily marked. The highly reticulated plants have been given the name *G. oblongifolia* var. *reticulata* Bovin, but the veining varies continuously between the nearly unmarked and the highly reticulated leaves, and a varietal name therefore has little value. The seed capsules, ovoid-ellipsoid, are held in a semi-erect position.

Menzies' rattlesnake plantain is unique among California's orchids in having leaves that persist for more than one year. The rosette of leaves precedes flowering by three years (Sheehan, 1992), and the leaves persist for at least one year after flowering. It is common to see this year's flowering growth accompanied by last year's dried flower stem, held in the center of still-fresh leaves. Near the upper limit of the species' elevational range, winter snows are frequent, accumulating to depths of over a meter. The leaves on the plants that will bloom the following spring survive beneath the snows, and the reticulated leaves peaking through the retreating snows in late spring are a promise of the blooms to come.

The plants propagate in part via an underground creeping rhizome with fibrous roots, which develops multiple leads each year. As the center rosette dies off, these new leads set up new groups of rosettes, eventually producing the massed colonies found in some areas. The plants also propagate by seed, producing rosettes in the fifth year after germination (Sheehan, 1992). There has been some progress in the asymbiotic germination (lack of compatible fungus) of *G. oblongifolia* seeds (Arditti et al., 1982).

The pollination biology of *G. oblongifolia* was described by Ackerman

(1975). Almost amazingly for such a small flower, the pollinators are queen bumblebees of *Bombus occidentalis* Greene. Ackerman observed that on newly opened flowers, the proximity of the column to the lip prevents the bee from probing very deeply into the flower. In the process of probing for nectar, the bee contacts the sticky viscidium, and as it departs it carries away pollinia. If a pollinia-laden bee visits another newly opened flower, the position of the column blocks access to the stigma, but as the flower ages, the column moves away from the lip, allowing pollinia on visiting bees to contact and adhere to the stigma. Individual flowers of *G. oblongifolia* last for two weeks, and seed capsules mature in six to eight weeks. Ackerman reported an average capsule set of 46.2% with average seed viability of 83.8%.

Distribution: *Goodyera oblongifolia* occurs across a large expanse of North America, from the southern portion of Alaska as far south as Mexico and as far east as the Gulf of St. Lawrence. The rattlesnake plantain grows in 26 counties in Northern California, and is nearly continuous along the coast north of Santa Clara and San Mateo Counties. Its southernmost location, in the Sierra Nevada, is in Tulare County. There is no record of rattlesnake plantain ever occurring in Southern California.

Habitat: *Goodyera oblongifolia* grows between sea level and 2130 meters elevation in California. In other parts of its range, it grows at even higher elevations: Luer (1975) reports it at 3050 meters in Arizona and New Mexico, and Long (1970) reports it between 2300 and 3050 meters in Colorado. It is interesting that its lowest elevation in Colorado is above its highest elevation in California. The most typical habitat for *G. oblongifolia* is in the humus of coniferous forests, in light to deep shade. Along the coast it grows in the sandy soil of beach pine forests, only a few meters away from active sand dunes. In the mountains it prefers forest duff, and in moist areas, *G. oblongifolia* will colonize and bloom on the tops of decaying logs. In the proper environment, colonies of hundreds of plants develop. The massed leaves make this one of the easiest orchids to find. It is visible from roads and trails, and occurs in campgrounds. The distinctive spires of the flower spikes are detectable from a considerable distance, which adds to the ease of finding it.

Blooming season: *Goodyera oblongifolia* blooms relatively late, but flowers over a long period. The earliest recorded bloom is in early May. More commonly, as in the Sierra Nevada, it first opens in late June and early July. It remains in bloom after most of our other orchids have faded. It is sometimes still in bloom in September in Sequoia National Park, and Ackerman (1975) reports flowering into October in Humboldt County. *Goodyera oblongifolia* frequently grows in the company of other orchids, such as the piperias, coral-roots, and lady's slippers. Usually, it is just spiking when they are blooming, and by the time *G. oblongifolia* opens, the others have long since formed seed capsules.

Conservation: The rattlesnake plantain is widespread and numerous. It occurs in the protected habitat of many State Parks and National Parks. The distinctive foliage of the plants, however, makes them easy prey for collectors. Watson (1992) reported an attempted theft of wild plants from Picture Rocks National Lakeshore in Michigan. Fortunately, the person attempting to remove the plants was apprehended, and they were replanted.

8. *Listera* R. Brown

In Aiton, Hortus Kewensis, ed. 2, 5: 201. 1813.
Etymology: The genus was named in honor of Martin Lister, an English botanist and physician.

Listera (liss′ - ter - ah) is a worldwide genus of 25 species. Eight of the species grow in the United States and Canada, three of these in California. Plants in this genus are commonly referred to as twayblades. The twayblades are characterized by a pair of opposing leaves situated midway along a slender stem. The stem is supported by a few fibrous roots. *Listera* includes one of the rarest orchids in California and one of the more common. Two of the three California species barely make it into the state, reaching their southern limits in our northernmost counties. They are among our most difficult orchids to find because of their small size and concealing habitat. Their intricate beauty will amaze and fascinate, but requires a magnifying glass to be appreciated.

There was much confusion about *Listera* in the last century, with all three of the species in California at one time or another being given the specific epithet *convallarioides*. Wiegand (1899) clarified the nomenclature with his revision of the genus.

Key to the California Species of *Listera*

1. Lip deeply forked for about half its length — *L. cordata*
1. Lip not forked:
   2. Lip abruptly contracted at base — *L. convallarioides*
   2. Lip narrowing gradually to base — *L. caurina*

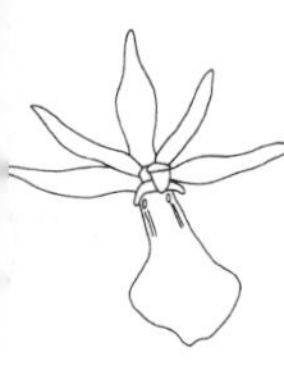

Listera caurina Piper

Erythea 6: 32. 1898.

Etymology: The epithet, from Latin, means "of the north wind," in reference to the area where the species occurs.

Synonymy:
Listera convallarioides Hooker, Flora Boreali-Americana: pl. 205. 1840.
Listera retusa Suksdorf, Deuts. Bot. Monats. 18: 155. 1900.
Ophrys caurina Rydberg, Bulletin Torrey Botanical Club 32: 610. 1905.

Common name: northwest twayblade.

Plate 18

Description: Large specimens of *Listera caurina* (kaw - ree′ - na) exceed 25 cm in height and bear upwards of 30 tiny green flowers, each about 6 × 11 mm. Plants as small as 8 cm tall can bloom. The two opposite leaves, arising roughly midway on the stem, are longer than wide, measuring approximately 4.75 × 3.0 cm on large plants. In size and appearance the light-green flowers of *L. caurina* resemble those of *L. convallarioides,* but are easily differentiated. *Listera caurina* has a shorter column, but differs primarily in the characteristics of the lip. From its well-spread tip, the lip narrows uniformly as it approaches the column, without the distinct step-down of *L. convallarioides.* The lip measures 4 mm at the widest part near the tip, and is about 6 mm long. *Listera caurina* is most easily identified by the pair of dark-green stripes running along the sides of the lip near the column. On some plants the green markings more closely resemble a set of dark-green eyes than stripes. On freshly opened flowers, the lanceolate sepals and linear petals are usually cupped forward over the column, but they reflex backward as the flower ages. A pair of minute horns arise where the lip joins the column. The spherical capsules are held in a semi-erect position.

Distribution: The northwest twayblade ranges from the northern part of California into Alaska, and the easternmost extent of it range is near Yellowstone National Park. In California it occurs in just Del Norte, Siskiyou, Humboldt, and Mendocino Counties.

Habitat: *Listera caurina* grows in the dry litter of mixed and coniferous forests at elevations between sea level and 1970 meters. It mixes in well with other small plants and often is very difficult to find. Many colonies consist of only a few plants, and even within the larger colonies the plants are widely scattered. In a few locations, *L. caurina* grows in damp to wet conditions on moss-covered rocks or in mossy soil. It occurs infrequently in campgrounds.

Blooming season: The northwest twayblade starts blooming in late April. As with many of our wild orchids, unseasonably warm weather can accelerate the end of the blooming period. During years with an unusually warm spring and early summer, such as 1992, *L. caurina* will finish blooming by early June in all but its highest locations. In most years, it is still in bloom in the high-elevation parts of its range in mid-July. Occasionally, *L. caurina* blooms with *L. cordata.* Less often, the northwest twayblade blooms with *L. convallarioides. Calypso bulbosa, Corallorhiza mertensiana, C. maculata,* and *Goodyera oblongifolia* also bloom in the same area as the northwest twayblade.

Conservation: Although it occurs in only four counties, *L. caurina* is widely scattered and grows in several of the State Parks within its range. Because of this growth pattern, the plant is fairly safe from threats, although destruction of habitat by logging continues.

Listera convallarioides (Swartz) Nuttall

Genera North American Plants 2: 191. 1818.

Etymology: *Convallarioides* means "like convallaria," or, "like lily of the valley," in reference to a supposed resemblance of the orchid to the popular lily of the valley.

Synonymy:
Epipactis convallarioides Swartz, Vet. Akad. Handl. Stockholm 21: 232. 1800.
Ophrys convallarioides (Swartz) Wright ex House, Bulletin Torrey Botanical Club 32: 380. 1905.
Bifolium convallarioides (Swartz) Nieuwland, American Midland Naturalist 3: 129. 1913.

Common name: broad-leaved twayblade.

Plate 19

Description: Plants of *Listera convallarioides* (kon - va - lahr - ee - oy′ - deez) reach 25 cm in height and carry over 25 small green flowers. The two opposite leaves, arising midway on the stem, each measure about 4.5 × 3.5 cm. The flowers are the largest of the genus in California. Elsewhere, reportedly, the flowers are sometimes marked with purple (Bingham, 1939), but those in California are all green. The translucent lip is the most prominent feature of the flower, measuring 1.0 cm long by 0.5 cm at its broadest expanse near the tip. The lip, notched slightly at the apex, necks down abruptly as it nears the column, forming a narrow handle, a characteristic that renders the plant readily recognizable as this species. The diminutive, lanceolate sepals and the linear petals are sharply reflexed back around the ovary, and the column curves prominently over the lip. (Flower shape and presentation suggest prehistoric birds.) Like other orchids, *L. convallarioides* has special adaptations to aid pollination (see Ramsey, 1950). A minute projection from the rostellum acts like a trigger. When the trigger is bumped by a visiting insect, the pollen masses are fired onto the intruder, which then transports them to the next flower visited. The capsules are nearly spherical.

Distribution: *Listera convallarioides* is widely distributed in North America, ranging from Southern California into Canada and across to New England. A finger of distribution reaches down through Idaho and into the Rocky Mountains. Within California the species occurs in 30 counties, and is by far the most common of our twayblades. It grows in many places in the Sierra Nevada, the San Bernardino Mountains, and the northern coastal mountains. The broad-leaved twayblade most likely also occurs in Amador County because the county shares habitat and proximity with counties where the broad-leaved twayblade is plentiful.

Habitat: *Listera convallarioides* is found in moist shady places between 760 and 2900 meters elevation, thus at much higher elevations than are the other two twayblades that occur in California. Most often, *L. convallarioides* grows hidden in mosses and grasses or under bushes on the edges of small streams. In mountain meadows it grows in the lee of a fallen log, or tucked inconspicuously in depressions in the fingers of the meadow as it intergrades into the forest. At other times it can be found on boggy hillsides under the cover of protecting ferns. Within the Sierra Nevada, *L. convallarioides* is frequently found in damp spots in sequoia groves or on moss-covered rocks alongside or even within mountain streams or hillside seeps. During years of high rainfall, the streamside plants are often growing in water, sometimes completely covered, until late in the season. In most of these habitats, fairly large and dense colonies often form, creating a nearly continuous ground cover.

Blooming season: *Listera convallarioides* is the last of the state's three twayblades to open, but its blooming season is one of the longest for California's orchids, stretching from late May to late August. High in the mountains, where it blooms near the end of summer, it is one of the last high-elevation orchids to open. Where its range overlaps that of *L. caurina,* the two twayblades sometimes bloom together. Other orchids that bloom nearby include *Cephalanthera austiniae, Corallorhiza maculata, C. striata, Cypripedium montanum, Platanthera dilatata,* and *P. sparsiflora.*

Conservation: Its wide distribution and large numbers make *L. convallarioides* relatively safe from threats, but as is the case with all of our orchids, individual colonies are continually at risk from both the activities of man and the vagaries of nature. For many years a small colony grew in Yosemite National Park on the bank of a small stream that meandered alongside a picnic area. The plants were in mosses nestled among the roots of a tree on the stream's edge, only 30 cm or so above the water. *Listera convallarioides* was one of seven orchid species within 25 meters of a picnic table at that site. A storm in the late 1980s toppled the tree into the stream, carrying the small colony of *L. convallarioides* with it, and the twayblades were washed away. Fortunately, *L. convallarioides* still thrives in massive colonies elsewhere in Yosemite.

Notes and comments: *Listera convallarioides* is one of the more difficult orchids to find in California. Because of its small size it lies effectively camouflaged among the protecting vegetation, and it tends to grow in out-of-the-way places. I usually find it serendipitously, while looking for some other orchid, rather than as the result of a conscious decision to look for the broad-leafed twayblade.

Listera cordata (Linnaeus) Brown

In Aiton, Hortus Kewensis, ed. 2, 5: 201. 1813.

Etymology: The epithet *cordata* derives from Latin, meaning heart-shaped, in reference to the shape of the leaves.

Synonymy:
Ophrys cordata Linnaeus, Species Plantarum: 946. 1753.
Epipactis convallarioides Bigelow, Florula Bostoniensis, ed. 2: 323. 1824.
Listera nephrophylla Rydberg, Memoirs New York Botanical Garden 1: 108. 1900.
Ophrys nephrophylla Rydberg, Bulletin Torrey Botanical Club 32: 610. 1905.
Bifolium cordatum Nieuwland, American Midland Naturalist 3: 129. 1913.
Listera cordata (Linnaeus) R. Brown var. *nephrophylla* (Rydberg) Hultén, Flora Aleutian Islands: 145. 1937.

Common name: heart-leaved twayblade.

Plate 20

Description: Blooming plants of *Listera cordata* (kor - dah′ - ta) range from under 6 cm to over 24 cm from the ground to the top of the flowers. The two opposite, heart-shaped leaves, arising midway on the stem, measure about 3 × 3 cm. Very rarely, a plant will have three leaves instead of the usual two, the third leaf appearing between the primary pair and the lowest flower. The largest plants carry up to 27 flowers, and the smallest always seem to carry at least three. The flower of the heart-leaved twayblade is unmistakable. The slightly ovate petals and sepals form a star over the lip. The deeply forked and spreading lip resembles legs, giving the flower an elflike appearance. Two hornlike processes project laterally near the base of the lip. The ovaries are swollen on even the freshest flowers, and the nearly spherical seed capsules mature and dehisce in a very short time, usually within five weeks of blooming. Seeds are sometimes released from lower capsules while the upper flowers are still functional (Stoutamire, 1964).

Some authorities recognize two varieties of *L. cordata* in the United States (see Luer, 1975; Correll, 1978). They differentiate between the typical *L. cordata* var. *cordata* and var. *nephrophylla* (Rydberg) Hultén on the basis of color, flower size, and leaf size. Var. *nephrophylla,* which occupies the far western limits of the species' range, on the average has slightly larger leaves and flowers. Luer and Correll maintain that flower shape is identical in the two varieties, but that the flowers of var. *nephrophylla* are always green, whereas those of var. *cordata* vary from green to red. Correll (1950) says the two varieties often grow together, with very few intermediates. Hitchcock (1969) does not recognize the two varieties, arguing that where the two proposed varieties grow together, flowers of intermediate colors are found, rendering the separation of taxa on the basis of flower color not feasible. Calder and Taylor (1968) agree with Hitchcock, saying that all morphological characteristics except leaf shape are shared between eastern and western plants. They present data showing that although leaf size and leaf shape overlap, the leaves of the western plants are broader than long, those of the eastern plants longer than broad. Nonetheless, they do not believe the distinction sufficient to maintain separate varieties.

The pollination mechanics of *L. cordata* were described by Ackerman and Mesler (1979). The flower possesses a trigger composed of three pressure-sensitive hairs at the tip of the rostellum. A droplet of glue is ejected onto a visiting insect that touches one of the hairs, and the pollinia are then released to fall on the glue. Once the pollinia have been removed, the spread rostellum remains as a protective cover over the stigma for about one day, after which it gradually lifts, again exposing the stigma.

Visiting insects are then able to deposit pollinia gathered elsewhere, thus completing the pollination process. This elaborate procedure virtually ensures cross-pollination, and demonstrates one of the many specializations of orchids. Ackerman and Mesler identified the most common pollinators as belonging to the fungus gnat families Mycetophilidae and Sciaridae.

Distribution: *Listera cordata* is the most widely distributed of the *Listera,* growing in much of the Northern Hemisphere. Conversely, it is the least common of the twayblades in California, occurring only in the northwest counties of Del Norte, Humboldt, and Mendocino. Because several colonies of *L. cordata* grow near the border of Humboldt and Trinity Counties, there is a fairly good chance the species may one day be found in Trinity County also.

Habitat: *Listera cordata* has a fairly limited elevation range in California; it is seldom seen below 40 meters, and rarely grows above 600 meters. In Colorado, however, it grows at much higher elevations: Long (1970) reports it above 3000 meters there. *Listera cordata* grows in very dry to fairly damp habitat, usually under redwood, oaks, madrone, or firs. The most common habitat is in dry conifer duff. In damper spots in the forest, plants grow out of moss and colonize the tops of decaying logs and tree stumps. The heart-leaved twayblade also grows on fairly steep hillsides or roadcuts, and less often grows in damp, but not wet, ditches or gullies. Most often, it occurs as scattered individuals or small groups of plants, but occasionally it forms large colonies. One colony in Humboldt County contained close to 1000 plants, and seemed equally distributed between those with pure-green flowers and those in various shades of pink.

Blooming season: *Listera cordata* is one of our earliest orchids, first blooming in the latter part of March. At the upper limits of its elevation range, it can remain in bloom until late June, but its peak period is in April and early May. By late May in most years, the flowers have faded and the seed capsules are maturing. *Listera cordata* blooms with *L. caurina, Calypso bulbosa,* and *Piperia candida. Goodyera oblongifolia* grows nearby, but blooms much later.

Conservation: Because of its rarity in its limited penetration into our state, *L. cordata* is included on the Watch List, List 4, in the fourth edition of the California Native Plant Society's *Inventory of Rare and Endangered*

Vascular Plants of California (Smith, Berg, et al., 1988). Inclusion on this list means that its "vulnerability or susceptibility to threat appears low at this time," but that populations should be monitored for changes in status.

Notes and comments: When first observing this species in the field, I supported the concept of two varieties (Coleman 1989d), but after additional field observations and study of herbarium specimens from across the United States and Europe, I came to agree with Hitchcock, and with Calder and Taylor, and no longer separate *L. cordata* into two varieties.

9. *Malaxis* Swartz

Prodromus Descriptionem Vegetabilium in Indian Occidentalem: 119. 1788.
Etymology: *Malaxis* is Greek for softening, in reference to the soft leaves on plants in this genus.

The genus *Malaxis* (ma - lax' - is) consists of over 200 members distributed worldwide. Nine species and varieties of *Malaxis* grow in the United States, but only one is found in California. *Malaxis* is most closely related to *Liparis*, a genus that does not occur in California but is found in the northeastern part of the United States and in Canada and in many other places around the world. *Malaxis* is distinguished from *Liparis* by having both a shorter column and pollinia with tapered ends. In addition, the lip of *Malaxis* is distinctly three-lobed, and the lateral lobes close partially around the column. Because of its limited distribution in California, *Malaxis* can be identified almost entirely by location. It can be readily distinguished from *L. convallarioides*, which grows nearby, by having a single leaf, instead of two, and a lip that is pointed, instead of broadly bilobed, at the apex. The flowers of *Malaxis* are so small, however, that identifying the distinguishing characters requires a microscope, and it is therefore best to rely on the overall structure of the plant in attempting identification in the field.

Malaxis monophyllos (Linnaeus) Swartz var. *brachypoda* (A. Gray) Morris and Eames

Our Wild Orchids: 358. 1929.

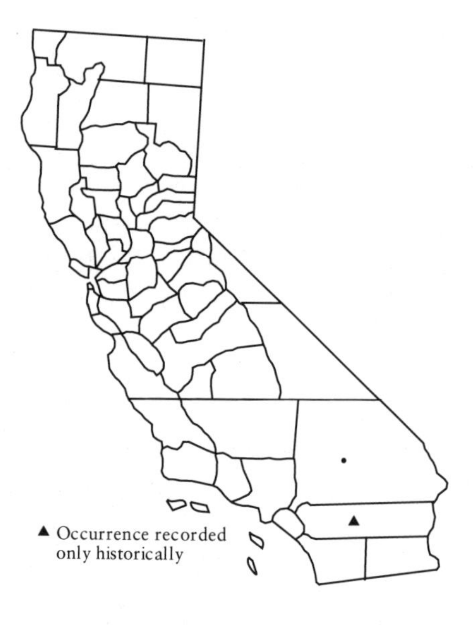

Etymology: The species epithet, *monophyllos,* is Greek for single leaf, in reference to the structure of the plant. The variety epithet, *brachypoda,* also Greek, means short-pediceled, in reference to the short pedicel of the flower.

Synonymy:
Microstylis brachypoda A. Gray, Annals Lyceum of Natural History New York 3: 228. 1835.
Malaxis brachypoda (A. Gray) Fernald, Rhodora 28: 176. 1926.

Common name: adder's tongue.

Plate 21

Description: *Malaxis monophyllos* (mon - oh - fill′ - ose) var. *brachypoda* (brack - i - poe′ - da) is the smallest wild orchid in California, standing barely 12 cm tall with a single basal leaf measuring about 5 × 2.5 cm. Small specimens are just 3 cm tall with a 2-cm leaf. Large specimens bear over 20 tiny pale-green to whitish flowers. The dorsal sepal is ovate-lanceolate, and the other two sepals are oblong-lanceolate. The filiform petals are much smaller than the sepals and reflexed sharply, sometimes touching each other behind the dorsal sepal. The lip is broadly triangular at the base and three-lobed, with the lateral lobes rolled forward. The lower half of the lip tapers sharply to a point and curves backwards. The ovoid capsules are held in a nearly erect position. The underground portion of the stem is thickened and cormlike.

Malaxis monophyllos var. *brachypoda* has gone by several names, including *Microstylis brachypoda* and *Malaxis brachypoda*. Luer (1975), Correll (1978), and Case (1987) all use *Malaxis monophyllos* var. *brachypoda,* the name established by Morris and Eames (1929). Morris and Eames separated the variety from typical *M. monophyllos,* which grows in Europe, on the basis of its resupinate flower, which they claim is the only feature distinguishing the two plants. Wilken and Jennings (in Hickman, ed., 1993) use the nomenclature *M. monophyllos* ssp. *brachypoda* rather than var. *brachypoda.*

Distribution: *Malaxis monophyllos* var. *brachypoda* is distributed across Canada and in much of New England and the Great Lakes region. The adder's tongue also occurs in three disjunct locations just outside its main range: in Illinois, in the Rocky Mountains of Colorado, and in Southern California. These isolated populations are perhaps relics of cooler, damper times when the adder's tongue may have been more widespread. *Malaxis monophyllos* var. *brachypoda* is of particular interest to us in California because for 42 years it had been our lost orchid. Historically, it was known from along the South Fork of the Santa Ana River in the San Bernardino Mountains of San Bernardino County, and near Tahquitz Valley in the San Jacinto Mountains of Riverside County. In 1989 it was rediscovered in San Bernardino County.

Habitat: The South Fork Meadows area, fully contained within the boundaries of the San Gorgonio Wilderness, stretches just over 2.5 kilometers between 2200 and 2800 meters elevation, averaging about 0.5 kilometer in breadth. Here, several large meadows and many small meadows and seeps are intermixed with forest. The adder's tongue prefers cool, damp areas in and on the edges of the meadows, in hillside seeps, and along wet river banks, often among grasses. Some of the plants grow in areas that are under water during periods of peak runoff. Other plants grow in an open meadow, with corn lilies providing the only protection from full sun. The orchids grow in short sedges and grasses right at the feet of the corn lilies. To find the plants it is necessary to get down on your hands and knees and spread the grasses.

Blooming season: The blooming season for *M. monophyllos* var. *brachypoda* stretches from early July to the latter part of August. The prime season is mid-July, although peak blooming varies slightly from year to year. Three other orchid species grow nearby: *Platanthera dilatata* var. *leucostachys* and *Listera convallarioides* grow in the meadows, and usually bloom at the same time as the adder's tongue; *Corallorhiza maculata* grows in the surrounding forest, but has usually faded when the others bloom.

Conservation: *Malaxis monophyllos* var. *brachypoda* should be treated as a threatened plant in California. Fortunately, its entire habitat in the San Bernardino Mountains is already protected because of its Wilderness Area status. Because there are at least three colonies scattered over the South Fork region, the probability that a single natural disaster such as a mudslide could destroy all the plants is slight. The ability of the plant to hide from searchers for 42 years indicates that it is safe from casual observation, and has a good chance for continued survival in California.

Notes and comments: The California Native Plant Society maintains an inventory of rare and endangered plants in the state, and *Malaxis monophyllos* var. *brachypoda* was included in the 1988 edition (Smith, Berg, et al., 1988) with the following note: "Several recent searches have not rediscovered CA occurrences."

All too often we tend to think of orchid conservation in the abstract, as something that needs to be pursued but requires infinitely more resources than we as individuals can supply. Opportunities, however, may exist right near our homes. They may be as simple as collecting habitat and population data on local orchids, or as challenging as reestablishing the presence of an apparently long-lost species. An effort of the latter sort rediscovered *Malaxis monophyllos* var. *brachypoda* in California.

The search to find *M. monophyllos* var. *brachypoda* was a cooperative effort. The Natural Diversity Data Base, part of the Non-game Heritage Program in the California Department of Fish and Game, conducts a program to learn more about the state's endangered plants. Each year Natural Diversity Data Base botanists select a group of rare and endangered plants and enlist volunteers for focused field surveys. The objective of the surveys is to increase the knowledge available for conservation work by collecting data on populations and habitat. After learning about their program, I wrote the Data Base and volunteered to search for *M. monophyllos* var. *brachypoda*. With assistance from state and federal agency personnel the search began in the San Bernardino Mountains in July 1989, and concluded successfully six weeks later with the discovery of a single plant. Additional searches conducted through 1992 revealed significantly more plants, and it is safe to assume that *M. monophyllos* var. *brachypoda* maintains a firm, but limited, foothold in San Bernardino County. Unfortunately, the same cannot be said of Riverside County; multiple searches there in 1989, 1990, and 1992 failed to turn up any plants.

10. *Piperia* Rydberg

Bulletin Torrey Botanical Club 28: 269. 1901.
Etymology: The genus was named in honor of C. V. Piper of the Agricultural Experiment Station at Pullman, Washington.

Piperia (pye - pur′ - ee - ah) has more species in California than does any other wild-orchid genus. The center of distribution for the genus is in fact in California, and some of the species are endemic to the state. Included in the genus are the four most recently described orchids in the state.

Taxonomic treatments of *Piperia* have varied over the years. Rydberg (1901) separated *Piperia* from the genus *Habenaria,* recognizing nine species. Ames (1910) included the plants within *Habenaria* and, of Rydberg's nine species, recognized only *P. elegans* and *P. unalascensis.* Correll (1978) also included the plants within *Habenaria,* and recognized only one species, lumping the various taxa as variants of *P. unalascensis.* Luer (1975), like Rydberg, separated *Piperia* from *Habenaria,* and recognized four species. Ackerman (1977) revised the genus, identifying four species and one subspecies. Thirteen years later, Morgan and Ackerman (1990) showed that the short-spurred plants usually lumped together under *P. unalascensis* consisted of at least four sibling species, two of which they described for the first time. Additionally, they elevated the subspecies Ackerman had identified in 1977 to a full species. Morgan and Glicenstein (1993) added one more species and a subspecies. I follow Ackerman, Morgan, and Glicenstein in recognizing ten *Piperia* species, all of which occur in California.

When Rydberg erected the genus *Piperia,* he differentiated it from *Pla-*

tanthera (Rydberg actually compared it to *Limnorchis,* but *Limnorchis* is included within *Platanthera* by most modern texts) on the basis of several features. Plants in both genera form underground tubers with a few fibrous roots, but in *Piperia* the tubers are rounded, and in *Platanthera* they are elongated. In *Piperia* most of the leaves are clustered near the base, and wither at anthesis, but in *Platanthera* the leaves are scattered along the stem, and remain green well past anthesis, usually until knocked down by frost. In *Piperia* the lateral sepals unite with the lip at their base, but in *Platanthera* the sepals are free. The lip has a medial ridge in *Piperia,* but is flat in *Platanthera.* Finally, the anther cells open laterally in *Piperia,* frontally in *Platanthera.*

The annual cycle of *Piperia* starts in the fall with the formation of new underground roots, one of which will later produce a replacement tuber. The mostly basal leaves appear in late fall to early spring, depending on elevation and timing of the rainy season. The two to six leaves mature through the spring, and in most species the flower spikes begin to arise in late spring and early summer. As the spike matures the leaves yellow and wither away. Often only a twisted brown trace of the leaves endures as the buds open. Each plant remains in flower for four to six weeks.

The flowers are "protandrous by age dependent lip movement" (Ackerman, 1977). On newly opened flowers, the lip is held tightly to the column, thus covering the entrance to the nectary. Visiting insects trying to probe for nectar contact the viscidium and remove pollen. As the flower ages, the lip moves downward, exposing the entrance to the nectary and allowing pollen deposition. By this stage, however, lip movement has effectively moved the viscidium out of reach, and the anther sacs have dried and shriveled, making further pollen removal nearly impossible. The effects of lip movement can be observed in the photographs of the species by comparing lip position on the upper and lower flowers.

Key to the California Species of *Piperia*

(developed with the assistance of Randall Morgan)

1. Sepals white with green midvein, the lip and petals white to pale green:
   2. Spur longer than lip, 7–13 mm:
      3. Plants slender, delicate, the scape mostly 1–3 mm diameter; spur

straight, stout for size of flower, mostly oriented horizontally; dorsal sepal usually projecting forward between the narrow, widely spreading petals; petals whitish with green midvein; strong, clove-like scent produced at night *P. transversa*

3. Plants relatively robust, the scape mostly over 3 mm diameter, sometimes thicker at base of raceme than at ground level; spur usually curved, decurrent along stem, often concealed; dorsal sepal more or less erect; petals whitish to pale green without well-defined midvein; scent strong at night, but not clove-like *P. elegans*

2. Spur about as long as lip, 2–6 mm:
    4. Petals and dorsal sepal projecting forward; flowers essentially white; raceme often more or less one-sided; stem bracts usually fewer than six; coniferous woods away from immediate coast, Santa Cruz County and north *P. candida*
    4. Petals and dorsal sepal more or less erect; flowers white with conspicuous green markings; raceme cylindrical, fairly dense; stem bracts usually more than six; narrow coastal endemics of central California:
        5. Petals falcate, often connivent at tips, green with broad white border on outer side only; scape attenuate toward tuber; flowers nearly scentless; pine woods and chaparral, northern Monterey County *P. yadonii*
        5. Petals more or less straight, spreading, green with white border all around or white with green center; scape slightly broadened toward tuber; flowers strongly cinnamon-scented; coastal headlands, Marin County *P. elegans* ssp. *decurtata*

1. Sepals and petals unmarked, green or yellow-green, often translucent but never white:
    6. Lip narrowly triangular-lanceolate; petals nearly linear, about 1 mm wide; spur 4–7 mm, curved, attenuate; flowers faintly lemon-scented at night; widespread, but rare *P. leptopetala*
    6. Lip triangular to deltoid, ovate, or lingulate; petals not linear; spur curved or straight, attenuate or rounded, 1–15 mm:
        7. Spur 1–6 mm in length:
            8. Lip more or less flat; scape base attenuate toward tuber;

A flower spike emerging in early spring, with its solitary basal leaf

A variant with white sepals and petals and a colored lip

A white variant with yellow highlights

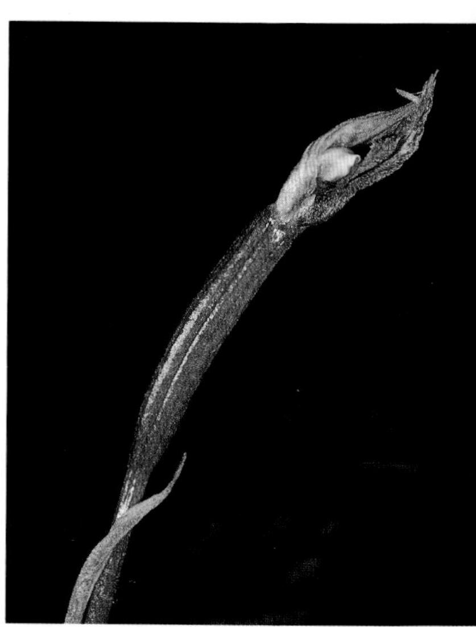

A seed capsule

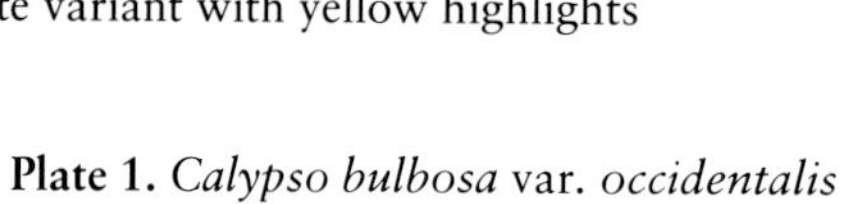

Plate 1. *Calypso bulbosa* var. *occidentalis*

A cluster of flowering stems

Hundreds of plants in one location

Seed capsules

The flowers

Plate 2. *Cephalanthera austiniae*

A cluster of plants

Inset: A typically colored flower

A variant with brightly spotted lips and bright reddish-purple stems, sometimes called *C. maculata* var. *punicia*

Another variant, sometimes called *C. maculata* var. *intermedia*

Plate 3. *Corallorhiza maculata,* 1

A lemon-yellow variant with white, unspotted lip, sometimes called C. *maculata* var. *flavida*

Another variant, with brownish petals and white lips

A narrow-lipped variant

Seed capsules

Plate 4. *Corallorhiza maculata,* 2

A group on the forest floor

Inset: A light-colored variant

Seed capsules

A typically colored flower; the lower sepals, as on this flower, are often folded back along the ovary

Plate 5. *Corallorhiza mertensiana*

Typically colored plants

A tan-colored variant, sometimes called *C. striata* var. *vreelandii*

A golden-yellow variant with red stripes

Plate 6. *Corallorhiza striata*

Spike of a typical plant

Two plants amidst the ground cover

The flowers of the same plant

Plate 7. *Corallorhiza trifida* var. *verna*

A large clump of plants in bloom

The flowers

Seed capsules; each flower node also supports a leaflike bract

Plate 8. *Cypripedium californicum*

Typically colored flowers

The nodding flower stems with their two opposed leaves

A pure-green variant

Plate 9. *Cypripedium fasciculatum,* 1

Flowers with pale-green pouches

Ten flowers in a tiny cluster

Seed capsules

Plate 10. *Cypripedium fasciculatum*, 2

Blooming in a dry habitat

A typical two-flowered stem

Plate 11. *Cypripedium montanum*, 1

A potential pollinator struggling to emerge from the basal end of the pouch

A seed capsule

A flower with two full lateral sepals instead of a synsepal

Plate 12. *Cypripedium montanum*, 2

Growing in a desert seep next to yucca plants

Inset: A nearly anthocyanin-free varian

A red-leaved variant from Sonoma County

Plate 13. *Epipactis gigantea*, 1

A typically colored flower

A syrphid fly emerging with pollen

An instance of peloria, a genetic aberration; all three petals have the shape of nearly perfect lips

Plate 14. *Epipactis gigantea*, 2

Plants in typical habitat

White plants, often called *E. helleborine* forma *monotropoides*

A variant with bright-pink flowers

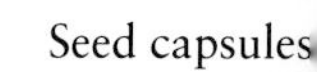

Seed capsules

Plate 15. *Epipactis helleborine*

Plant with minimal reticulation in the leaves

Plant with strong midvein reticulation

Plant with heavy reticulation

Plate 16. *Goodyera oblongifolia,* 1

The typical one-sided spike

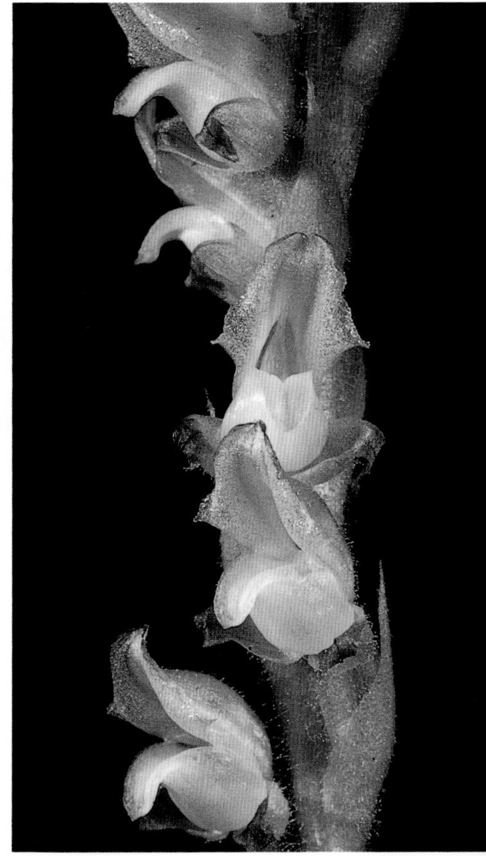

The flowers

Plate 17. *Goodyera oblongifolia,* 2

[T]he typical flower

Plant growing in pine litter, with its two opposite leaves midway up the stem

Seed capsules

Plate 18. *Listera caurina*

The typical flower

Plant in bloom, with its two opposite leaves

Seed capsules

Plants massed beneath ferns

Plate 19. *Listera convallarioides*

The opposite leaves, midway up the stem

The typical green flowers

Variant with reddish flowers

Plate 20. *Listera cordata*

The flower spike

Plant in bloom, with its single basal leaf

Seed capsules

Plate 21. *Malaxis monophyllos* var. *brachypoda*

The spike, not yet in bloom, arising between the two basal leaves

Seed capsules

The flowers

Plate 22. *Piperia candida*

The minute flowers

The grasslike leaves, with emerging spikes

Flowering spike

Seed capsules

Plate 23. *Piperia colemanii*

Seed capsules

Masses of plants, blooming along the coast

Plant in typical chaparral habitat

The flowers

Plate 24. *Piperia cooperi*

Plants in coastal habitat

Spike in full bloom

Plate 25. *Piperia elegans*, 1

The variant called *P. elegans* ssp. *decurtata*

The flowers of an inland plant

The flowers of a coastal plant

The leaves

Plate 26. *Piperia elegans*, 2

The leaves

In bloom in the chaparral

The flowers

Seed capsules

Plate 27. *Piperia elongata*

The leaves

The flower spike

The flowers

Plate 28. *Piperia leptopetala*

The leaves

The flower spike of an inland plant

The flowers of a coastal plant

Plate 29. *Piperia michaelii*

Plants in bloom
in Yosemite Valley

Seed capsules

The flowers

The leaves

Plate 30. *Piperia transversa*

Two spikes at the base of a tree in Siskiyou County

The flowers

Plate 31. *Piperia unalascensis*

The leaves

A laxly flowered spike

A densely flowered spike

Spike alongside a fallen tree in Monterey County

Plate 32. *Piperia yadonii*

Inset: Flowers with widely dilated lips   Spike amid a dense clump of *Epipactis gigantea*

Flowers with nearly linear lips

Seed capsules

Plate 33. *Platanthera dilatata* var. *leucostachys*

Plant growing on a dry hillside

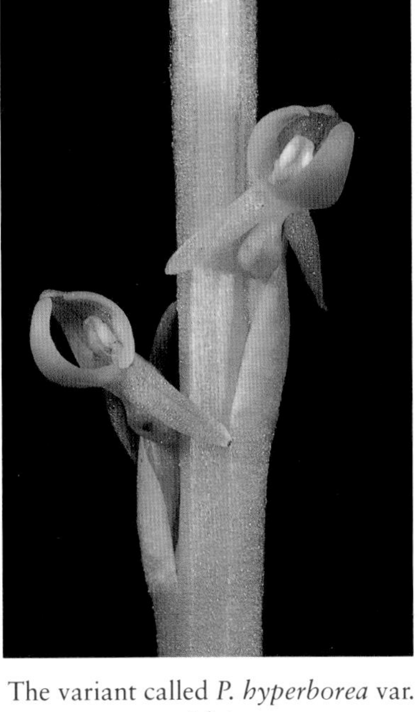

The variant called *P. hyperborea* var. *purpurascens*, with its saccate spurs

The flowers

Plate 34. *Platanthera hyperborea*

Plant growing on a wet stream bank

Seed capsule

The flowers

Plate 35. *Platanthera sparsiflora*

A flowering spike

The flowers

Seed capsules

Plate 36. *Platanthera stricta*

Plants growing on a wet, rocky bank

The spiral inflorescence

The flowers

Plate 37. *Spiranthes porrifolia*

Plant growing in a wet meadow

The twisted inflorescence

The flowers

Plant growing in a dry coastal habitat

Plate 38. *Spiranthes romanzoffiana*

P. ×*estesii*

P. ×*lassenii*

P. ×*media*

Undescribed *Spiranthes*, potentially a hybrid or new species

Plate 39. *Platanthera* hybrids and *Spiranthes* sp.

Piperia candida Morgan and Ackerman

Lindleyana 5(4): 205–211. 1990.

Etymology: The epithet *candida,* from Latin *candidus,* white, recognizes the species as the most completely white-flowered member of the genus.

Synonymy:
Piperia unalascensis (Sprengel) Rydberg, Bulletin Torrey Botanical Club 28: 270. 1901 in part.

Common names: pearl orchid, slender white piperia.

Plate 22

Description: *Piperia candida* (kan′ - di - da) had been considered a white form of *P. unalascensis* until elevated to species status by Morgan and Ackerman (1990). The tallest plants are about 60 cm, but some bloom at under 20 cm. The white flowers, sometimes numbering over 100 per inflorescence, have a honeylike scent with some harsh components reminiscent of urea. There is usually a green midvein on the dorsal sepal and some traces of green in the other flower parts. The lip is triangular and strongly downcurved. The lower sepals are somewhat twisted. The short, tapered spur is curved and of about the same length as the lip. The rather densely flowered inflorescence often carries all the flowers on one side of the stem. The two or three basal leaves, approximately 3 × 10 cm, angle up slightly from the ground. The largest leaf is about 3.5 × 15 cm. A fourth leaf, if present, is bractlike and arises partway up the stem. During high-rainfall years or in the damper parts of its habitat, the leaves of *P. candida* will often persist well into anthesis. The ellipsoid capsules are held upright along the stem.

Distribution: *Piperia candida* grows only in the Pacific Northwest, ranging from the middle of California into the southernmost tip of Alaska. Within California it occurs in just eight counties: Santa Cruz, San Mateo, Sonoma, Mendocino, Humboldt, Del Norte, Siskiyou, and Trinity. Even though the species is widely distributed, it is uncommon, and seldom forms large colonies as some of the other piperias do.

Habitat: *Piperia candida* inhabits coniferous and mixed evergreen forests from near sea level to about 1200 meters elevation. It grows in forest openings in nearly full sun, in the denser shade of heavily forested areas, and sometimes under shrubs. In open regions its roots penetrate the soil, but it also roots shallowly in forest litter and in mosses. It grows on gravel bars in rivers, as well as on flat terrain, on steep hillsides, and on the banks of drainages.

Blooming season: The slender white piperia blooms close to the middle of the wild-orchid season. Flowering begins in late May and lasts into early September. As with most of our orchids, the blooming time correlates closely with elevation: the higher the elevation, the later the blooming season. *Piperia candida* blooms with several other orchids, including *Calypso bulbosa, Corallorhiza maculata, Cypripedium montanum,* and *C. fasciculatum. Piperia elongata, P. transversa, P. unalascensis,* and *Goodyera oblongifolia* bloom at slightly different seasons, but grow in the same area as *P. candida.*

Conservation: Because of its wide distribution, *P. candida* is probably safe from threats. Many of its colonies are in State Parks or Wilderness Areas, and therefore enjoy protected habitat. Colonies outside these protected areas, however, are subject to loss from logging operations.

Piperia colemanii Morgan and Glicenstein

Lindleyana 8(2): 89. 1993.

Etymology: This species was named in honor of R. A. Coleman, who first pointed out the differences between *P. colemanii* and *P. unalascensis.*

Synonymy:
Piperia unalascensis (Sprengel) Rydberg, Bulletin Torrey Botanical Club 28: 270. 1901 in part.

Common name: none.

Plate 23

Description: *Piperia colemanii* (kole′ - men - ee - eye) reaches just over 50 cm tall, and bears as many as 100 tiny green flowers. The basal leaves, usually two or three, are conduplicate, slender, and grasslike, and angle up slightly from the ground. The leaves are most often withered at anthesis, but occasionally some traces of green are still present in the leaves when the plant is in flower. The lateral sepals are strongly decurved, and the petals are slightly decurved. The dorsal sepal projects forward between the petals. The triangular-lanceolate lip curves sharply upward, sometimes touching the dorsal sepal. The curved spur is clavate, decurrent, and very short, usually under 2 mm. *Piperia colemanii* is very similar to the more common *P. unalascensis,* but differs in several respects, the most noticeable being spur length: The spur of *P. colemanii* is always shorter than the lip, whereas the spur of *P. unalascensis* is about twice as long, equaling or exceeding the length of the lip. The leaves of *P. unalascensis* are broader, flat, and spreading, and the plant has a musky scent, most pronounced at night or evening. The lip of *P. colemanii* is more upcurved and more triangular-lanceolate than that of *P. unalascensis,* and the flower is scentless. The ellipsoid capsules of *P. colemanii* (like those of *P. unalascensis*) are held erect.

Distribution: *Piperia colemanii,* a California endemic, grows in a narrow strip stretching from the southern Sierra Nevada in Fresno County to the Cascade Mountains in Siskiyou County. There is a disjunct location in the North Coast ranges in Colusa County. The species occurs in just 11 counties within this range. The type specimen was collected by Abrams in Mariposa County.

Habitat: *Piperia colemanii* occurs primarily in open coniferous forests and chaparral between 1200 and 2300 meters elevation. It has a preference for very sandy soils, and Morgan and Glicenstein (1993) report it growing in deep, powdery sand. It also grows on the disturbed portions of roadbeds, and therefore is sometimes visible to passing motorists.

Blooming season: The blooming season for *P. colemanii* is relatively short, but that impression may simply be due to lack of data. The first flowers open in late June, and blooming extends into early August. The best time to look for it is in early July. Its blooming season overlaps with that of both *P. transversa* and *P. unalascensis,* both of which sometimes bloom in the same area. It also blooms with *Corallorhiza maculata, C. striata,* and *Cypripedium montanum.*

Conservation: *Piperia colemanii* occurs infrequently throughout its range. Morgan and Glicenstein report only 19 known locations, some of them within the boundaries of Yosemite National Park and therefore relatively protected. Most colonies, however, lie outside the protection afforded by State Parks and National Parks, and are subject to habitat loss due to logging. *Piperia colemanii* is rare and merits further study to determine if it warrants protection as an endangered plant.

Notes and comments: The characteristics of the lip and spur first drew my detailed attention to *P. colemanii.* I had seen the plants many times before, but always considered them to be *P. unalascensis* until finding both growing along the same trail. The plants were separated by about a kilometer, and I went back and forth between them, trying to determine why they looked different. After deciding the differences deserved further study, I referred my observations to Randall Morgan. He and Leon Glicenstein conducted more research into the plants, culminating their work with the description of *P. colemanii* (Morgan and Glicenstein, 1993).

Piperia cooperi (S. Watson) Rydberg

Bulletin Torrey Botanical Club 28: 270. 1901.

Etymology: Sereno Watson named the species in honor of J. G. Cooper, who discovered it.

Synonymy:
Habenaria cooperi S. Watson, Proceedings American Academy of Arts and Science 12: 276. 1877.

Common names: chaparral orchid, Cooper's stout-spire orchid.

Plate 24

Description: Many authors include *P. cooperi* (koop′ - er - eye) with *P. unalascensis* (Sprengel) Rydberg, but the plants are sufficiently different to maintain Rydberg's classification. The largest flower stems exceed 90 cm in height and carry over 100 blossoms. The tiny green, squat flowers have a broadly triangular or deltoid lip, hastate at the base. The short stubby spur, thick and clavate, is about as long as the lip. The dorsal sepal is ovate, and the lateral sepals and petals are oblong-lanceolate. Except on an occasional plant growing in damp habitat the two to four basal leaves are usually faded at anthesis. The largest leaf can be as long as 20 cm and as wide as 2 cm. The honeylike scent of *P. cooperi* is best detected at night, which suggests pollination by night-flying insects. The ellipsoid capsules are held erect.

Distribution: *Piperia cooperi* grows at the southern limit of the piperias, ranging from Southern California into Baja California, with its northern terminus in eastern Ventura County. Within California it occurs in only six counties—Ventura, Los Angeles, Orange, San Diego, San Bernardino, and Riverside—and on Santa Catalina Island. Throughout most of its range, *P. cooperi* is fairly rare, usually occurring in widely scattered colonies of only a few plants. Infrequently, it occurs in large colonies exceeding 50 plants scattered over several acres, and in 1992 an unusually large colony, consisting of thousands of plants, was discovered on U.S. Navy property on Pt. Loma in San Diego County. Watson (1876) described *P. cooperi* on the basis of specimens collected from clay hills near San Diego.

Habitat: *Piperia cooperi* sometimes grows on exposed ridges near the tops of chaparral-covered foothills and on mesa headlands near the ocean. It also grows on the lower slopes of some of the higher coastal mountains, but does not venture much above 900 meters. Having once seen our brown mountains, most summer visitors to Southern California would never believe that wild orchids could inhabit such barren landscape. The hills, however, are not always dry and brown. Southern California can be thought of as having two seasons, wet and dry. The wet, or rainy, season usually starts in November and lasts until April, and results in about 35 cm of rain each year, with 10 cm or more additional rain in the hills. In the dry season, from May to October, little if any rain falls. The hillsides that are a lush green in the winter and early spring start to turn brown by late spring, and remain brown until the rains start again. In this environment, most streams run only during the wet season. The streams that do flow year-round are usually a mere trickle by fall. *Piperia cooperi* grows in this harsh environment, in flat to fairly steep areas of the chaparral. It grows in full sun or under yuccas, manzanita (*Arctostaphylos* species), sage (*Salvia* species), or chamise (*Adenostoma* species). On exposed bluffs near the coast it grows in patches of non-native ice plant.

Blooming season: The leaves of *P. cooperi* emerge as early as mid-November from tubers buried 2 to 10 cm deep. The leaves mature in February and March, when the flower spike begins to elongate, and are usually totally brown by the time the flowers start to open in late March. Flowering lasts into early June, but the best displays usually occur between late April and early May. *Piperia elongata,* which grows in the same region as *P. cooperi,* blooms about a month later in the areas of overlap. *Epipactis gigantea* will be blooming along streams in the canyon bottoms at the same time *P. cooperi* is blooming high on the dry canyon walls.

Conservation: The habitat of *P. cooperi,* along with that of other wild orchids in Southern California, is under ever increasing pressure from urbanization. Many of its historical locations have been lost to development, and more colonies are undoubtedly destroyed each year. *Piperia cooperi* is also occasionally prey for those who insist on digging up our wild treasures. For many years three plants grew under a protecting sagebrush bush in the Santa Monica Mountains. They survived the long drought that began in 1987 and bloomed again in 1992. In early 1993, however, only one plant remained, growing near the base of the sage. Holes left by the diggers showed where the other two had been. Fortunately, many colonies of *P. cooperi* are protected within parks and reserves and are therefore safe from developers—and, we hope, from diggers. The rarity of *P. cooperi* is beginning to be recognized. The large colony on Pt. Loma is on property controlled by the Naval Command, Control and Ocean Surveillance Center, which has taken steps to protect the plants.

Notes and comments: Wildfires are a natural phenomenon in Southern California, and are as predictable as the dry months of summer. The wildfires that darken thousands of acres, claiming buildings and, unfortunately, lives result in some of the sightings of *P. cooperi.* Notes with many herbarium collections say that the plants were found in recently burned areas. Wildflower books and floras such as Raves, Thompson, and Prigge (1986), if they refer to *P. cooperi* at all, say that it is usually found after a fire. Plants found in such circumstances, however, did not populate the area after the fire, but rather had been growing under the chaparral, and merely survived the fire. The chaparral orchid goes dormant long before the start of the fire season, which may be one reason it lives through the fires. With the loss of the chaparral cover, the orchids bloom in the open the following spring, causing the increase in sightings.

Piperia elegans (Lindley) Rydberg

Bulletin Torrey Botanical Club 28: 270. 1901.

Etymology: The epithet *elegans* is Latin for elegant, in reference to the plant's comely inflorescence.

Synonymy:
Platanthera elegans Lindley, The Genera and Species of Orchidaceous Plants: 285. 1835.
Habenaria elegans (Lindley) Bolander, A Catalogue of the Plants Growing in the Vicinity of San Francisco: 29. 1870.
Habenaria maritima Greene non Rafinesque, Pittonia 2: 298. 1892.
Montolivaea elegans (Lindley) Rydberg non Reichenbach fil., Memoirs New York Botanical Garden 1: 106. 1900.
Piperia maritima Rydberg, Bulletin Torrey Botanical Club 28: 641. 1901.
Piperia multiflora Rydberg, Bulletin Torrey Botanical Club 28: 638. 1901.
Habenaria multiflora (Rydberg) Blankenship, Montana College of Agriculture Science Studies, Botany 1: 45. 1905.
Habenaria elegans (Lindley) Bolander var. *maritima* (Greene) Ames, Orchidaceae fascicle 4: 113. 1910.
Habenaria elegans (Lindley) Bolander var. *multiflora* (Rydberg) Peck, A Manual of the Higher Plants of Oregon: 219. 1941.
Habenaria unalascensis (Sprengel) S. Watson var. *maritima* (Greene)

Correll, Leaflets of Western Botany 4: 260. 1943.
Habenaria unalascensis (Sprengel) S. Watson ssp. *maritima* (Greene) Calder and Taylor, Canadian Journal of Botany 43: 1393. 1965.
Platanthera unalascensis (Sprengel) Kurtz ssp. *maritima* (Rydberg) DeFilipps, American Orchid Society Bulletin 44: 405. 1975.

Common names: coast piperia, elegant piperia.

Plates 25 and 26

Description: For much of this century, *Piperia elegans* (el′ - e - ganz) was more commonly known as *P. maritima* Rydberg, but Ackerman (1977) found that the specific epithet *elegans* held precedence. Plants of *P. elegans* can reach over 80 cm tall and support a densely flowered raceme. Plants in very exposed coastal sites, however, bloom at as little as 7 cm. More than 100 flowers surround the stem on the best specimens. Obtaining an accurate count of the flowers on a single raceme is difficult because they are packed so tightly together. The predominantly white flowers are the largest in the genus. Each can measure up to 1.3 cm across the sepals, with a spur up to 1.3 cm long. The dorsal sepal is ovate, the lateral sepals ovate-elliptic, and the sepals have a conspicuous green stripe down the middle. The petals vary somewhat from ovate-lanceolate to oblong-lanceolate. The lip is lanceolate and dilated at the base, with a prominent midvein. The filiform and decurrent spur is not obvious at first glance because of the density of the inflorescence. The flower stem, which is often thicker near the flowers than at the base, has many bracts. Plants along the coast have denser inflorescences and more bracts than those farther inland. The two or three basal leaves are large for this genus, nearly 3 × 30 cm. The coast piperia is pollinated by the noctuid moth *Chrysaspidea nichollae* and by an *Autographa* moth species (Ackerman, 1977). The ellipsoid capsules are held upright.

Morgan and Glicenstein (1993) identified a subspecies of *P. elegans* they called *P. elegans* ssp. *decurtata* Morgan and Glicenstein. They differentiated it on the basis of its short spur, about 5 mm, and the cinnamon-candy scent of the flowers. It is also consistently smaller in stature than its cousin, and is more strongly bicolored in white and green.

Distribution: *Piperia elegans* is distributed along the Pacific Coast as far north as British Columbia. In California it occurs in 17 primarily coastal counties from Santa Barbara to Del Norte. To judge from its distribution pattern and habitat preference, *P. elegans* may also occur in Solano County, but has not been documented there. *Piperia elegans* ssp. *decurtata* is much less widely distributed. It is endemic to Marin County, and grows only in a narrow band along the coast.

Habitat: *Piperia elegans* grows in a variety of habitats at elevations between sea level and 1000 meters. Over 50 different vegetative types, distinguished by the characteristics of the dominant plants, have been identified within California. The vegetative type known as coastal scrub (see Barbour and Major, 1988), consisting of evergreen shrubs less than 2 meters tall, grows along much of the central coast. Representative plants of the scrub include *Baccharis pilularis, Eriophyllum staechadifolium, Gaultheria shallon, Lupinus varicolor,* and *Rubus vitifolius.* The scrub provides cover for smaller herbaceous plants, including *P. elegans.* Totally unexpected, and often overlooked among the scrub, the orchids sometimes grow within reach of the salt spray from the ocean. On some of the more exposed bluffs, *P. elegans* grows protected by poison oak or among grasses and irises. Additional plants grow within the nearby coniferous forest, and occasional colonies venture inland for many kilometers. Its most picturesque habitat is on coastal bluffs and headlands, where it grows right up to the edge of cliffs and on the steep sides of canyons leading down to the surf. Some herbarium specimens were in fact collected from offshore rocks. Flowering plants appear like white candles sticking above the surrounding scrub, but sometimes the coast piperia grows in full sun. Usually, at least the leaves are protected by low-growing shrubs, but sometimes the entire inflorescence is concealed under loosely structured bushes. The coast piperia grows on roadcuts and in the iceplant prevalent along the coastal highways. Perhaps its strangest habitat is the non-native eucalyptus groves common along the central coast. *Piperia elegans* ssp. *decurtata* grows in mostly full sun on windswept coastal bluffs along the coast. It grows among grasses and low-growing native shrubs.

Blooming season: *Piperia elegans* starts to bloom about halfway through the wild-orchid season. The first flowers open in mid-May in the inland parts of the range, followed shortly by coastal plants near the southern range limits. Flowering extends into late September in the central-coast and northern counties. Where their ranges overlap, both *P. transversa* and *P. michaelii* bloom at the same time as the coast piperia. Along the central and northern coast, *Spiranthes romanzoffiana* shares the same dry, coastal cliffs and blooms at the same time.

Conservation: When considered across its entire range, the coast piperia is not threatened, because of its wide range and diversity of habitat, and in major portions of its range it grows within the protection of State Parks and State Beaches. Its favorite habitat, however, on the coastal bluffs, is also favored by developers, and colonies of *P. elegans* will continue to be lost as the coast is developed. *Piperia elegans* ssp. *decurtata* should be considered threatened because of its narrow endemism and its proximity to heavily visited areas. Morgan and Glicenstein estimate the subspecies' entire population at less than a few hundred plants, and report that it is often eaten by deer before its seeds can mature, which poses yet another threat to its existence.

Piperia elongata Rydberg

Bulletin Torrey Botanical Club 28: 270. 1901.

Etymology: The epithet, derived from Latin, refers to the elongated nature of the spur when compared with that of *P. unalascensis*.

Synonymy:
Habenaria elegans (Lindley) Bolander var. *elata* Jepson, A Flora of California: 330. 1921.
Habenaria unalascensis (Sprengel) S. Watson var. *elata* (Jepson) Correll, Leaflets of Western Botany 3: 246. 1943.
Habenaria unalascensis (Sprengel) S. Watson ssp. *elata* (Jepson) Calder and Taylor, Canadian Journal of Botany 43: 1393. 1965.
Piperia elegans (Lindley) Rydberg var. *elata* (Jepson) Luer, Native Orchids of the United States and Canada: 167. 1975.

Common names: chaparral orchid, wood rein-orchid.

Plate 27

Description: Flowering plants of *Piperia elongata* (ee - long - gay′ - ta) can reach nearly 1 meter tall with over 100 flowers. The inflorescence is usually loosely flowered, but sometimes is very densely flowered. The uniformly green flowers are easily recognized by the long spur and triangular lip. The spur, 0.8–1.5 cm long, is usually curved below the flower and held parallel to the ovary. The lip, usually about 4 mm long, is triangular in outline and slightly dilated at the base, with a raised swelling in the center. The dorsal sepal is ovate-elliptic, the lateral sepals oblong-lanceolate. The petals are ovate. The two to four basal leaves are typically dry at flowering. On very large plants there may be several leaflike bracts along the lower stem. *Piperia elongata* is the most deeply rooted of the genus, the tubers lying up to 12 cm deep in the soil. The elongated ellipsoid capsules are held upright along the stem.

Distribution: *Piperia elongata* is one of the more widely distributed piperias in California, occurring in 34 counties, from San Diego County to Del Norte County, and on the Channel Islands. It is a fairly widespread species on the west coast, and extends northward into Canada and eastward to Idaho and western Montana.

Habitat: *Piperia elongata* grows in a variety of habitats up to elevations of 2100 meters. In Southern California it occurs in the same chaparral community as *P. cooperi,* in very dry conditions. It seeks the protection of the chaparral shrubs on the side of a seasonal drainage, or at considerable distances from the water on the slopes of drainages supporting year-round streams. Farther north, it occurs in mixed coniferous or oak forest under pines, dogwood, madrone, and firs, occasionally growing near streams and frequently appearing on roadcuts. In all these habitats it will grow either in the open or in areas with heavy competition from other plants. In favorable conditions, large colonies will form; a roadcut in Mendocino County supported 61 blooming plants in an area of 2 square meters.

Blooming season: *Piperia elongata* starts blooming in early May. In the Southern California foothills it is usually in peak bloom by late May or early June, and by July it has finished blooming. In the northern parts of its range, and higher in the mountains, it does not start blooming until late June, and during some years will still be in bloom in early September. The usual pattern for our orchids is for the lower-elevation plants to bloom before those higher in the mountains. The pattern for *P. elongata* is for inland plants at the same latitude to bloom ahead of plants nearer the coast, even though the coastal plants are at much lower elevation. For example, in late June 1992 a colony at 1200 meters elevation was in full bloom, while a colony located nearer the coast at the same latitude, but at about 200 meters, had not started to open. *Piperia elongata* blooms at the same time and in the same area as *P. transversa,* but usually opens slightly later. *Piperia candida* and *Cypripedium fasciculatum* grow in the same area, but finish blooming before *P. elongata* opens.

Conservation: In Southern California, the habitat of *P. elongata,* along with that of *P. cooperi,* is under siege from relentless development. Each year more of the chaparral, and hence more of the orchids, is destroyed by development. Many locations, however, are protected within the Santa Monica Mountains National Recreation Area. Portions of *P. elongata*'s habitat in the northern counties are protected within State Parks and National Parks.

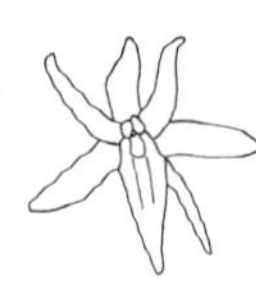

Piperia leptopetala Rydberg

Bulletin Torrey Botanical Club 28: 270. 1901.

Etymology: The epithet *leptopetala* means "thin, delicate petals."

Synonymy: none.

Common names: lace orchid, thin wood rein-orchid.

Plate 28

Description: The laxly flowered plants of *P. leptopetala* (lep - toe - pett′ - la) are delicate and slender, reaching 40 cm in height. The pale or translucent green flowers have a much more open appearance than do those of other piperias, appearing almost feathery. The sepals and two petals are linear-lanceolate. The lip is more lanceolate than are the sepals and petals, and the usually curving spur is as long as the lip to about twice as long. (Because of its variable spur length, *P. leptopetala* is often confused with *P. elongata* or *P. unalascensis.*) At night the flower has a faint, lemonlike aroma. The ellipsoid capsules are held upright along the stem.

Distribution: *Piperia leptopetala* is endemic to California, and is widely distributed but rare throughout its range. It occurs in 16 counties from San Diego to Del Norte, and reaches as far inland as Nevada County. Its center of distribution is in the central-coast region, but it is missing even in some of those counties.

Habitat: *Piperia leptopetala* prefers the same habitat as do many of the other inland piperias. It grows primarily in open mixed or coniferous forest, in light to moderate shade, between just below 400 to just over 2100 meters. Plants inhabit both fairly steep hillsides and relatively flat terrain. Like several of our other orchids, especially other piperias, *P. leptopetala* grows in stands of poison oak throughout much of its range.

Blooming season: *Piperia leptopetala* has a relatively short blooming season. Flowering begins in mid-May and has finished by the end of July. Peak blooming occurs in early June. In the coastal ranges, *P. leptopetala* blooms at the same time and in the same area as *P. michaelii, Epipactis gigantea,* and *Cypripedium californicum.* It usually blooms just ahead of *P. transversa* and just after *Corallorhiza maculata* where their ranges overlap.

Conservation: Because its distribution and growth patterns make it susceptible to loss of range, *Piperia leptopetala* should be considered a threatened plant. For example, in 12 of the 16 counties in which it occurs, there is only one record of it, and most colonies consist of only a few plants, usually less than 10, with a maximum of around 30. Because of its few, relatively tiny, colonies, a single logging operation or natural disaster could eliminate *P. leptopetala* from an entire county. At least one colony, however, exists in the relative protection of a State Park.

Notes and comments: In the spring and early summer, orchid hunters are rewarded with many beautiful flowers. Populations of biting insects however, peak at about the same time the orchids do. Mosquitoes are controllable with generous and frequent applications of repellent, though you quickly learn to expect at least a few bites. The flies are a different story, because repellents will not keep them completely at bay. No area of the body is safe, because they can bite even through clothing. The number of flies is at a maximum at the same time and elevation at which *P. leptopetala* blooms. As more than one field photographer has experienced, the flies bite their hardest precisely when you are about to take a picture. Just brace yourself for the bites and consider the welts part of the price for seeing the orchids.

Piperia michaelii (E. Greene) Rydberg

Bulletin Torrey Botanical Club 28: 270. 1901.

Etymology: This species is named in honor of G. W. Michael, who made the original discovery.

Synonymy:
Habenaria michaelii Greene, Bulletin California Academy of Science 1: 282. 1885.
Piperia elongata Rydberg ssp. *michaelii* (Greene) Ackerman, Botanical Journal Linnean Society 75: 266. 1977.

Common name: Michael's rein-orchid.

Plate 29

Description: Blooming plants of *Piperia michaelii* (my - kill′ - ee - eye) are typically 20–50 cm tall and carry from 20 to over 60 flowers, each measuring about 1.0 × 0.6 cm tip to tip. The monochromatic green to yellow-green flower is distinguished from that of *P. elongata* by the shape of its lip. The lip is widely triangular with slightly rounded edges, and the sepals and petals are ovate and triangular-ovate. The decurrent spur is much longer than the lip, but usually under 1.4 cm long. Plant structure—a very full raceme tops a many-bracted stem—bears some resemblance to that of *P. elegans*. The flowers, however, are green rather than white, the lip shape differs, the plants are usually shorter and less robust, and the cone of flowers is neither as large nor as dense as that of *P. elegans*. The stem, moreover, usually tapers from the base rather than thickening as in *P. elegans*. The three or four basal leaves are usually faded at anthesis, but may persist in shaded conditions. The ellipsoid capsules are held upright along the stem. Ackerman (1977) included *P. michaelii* as part of *P. elongata,* calling it *P. elongata* ssp. *michaelii.* Morgan and Ackerman (1990) revised that position, and now recognize *P. michaelii* as a separate species.

Distribution: *Piperia michaelii* is endemic to California, and has been documented in 20 counties. It grows mainly in a narrow band along the coast between Los Angeles and Marin Counties, but some plants have been found as far inland as the Sierra foothills and as far north as Humboldt County. Throughout that range, it is not as common as, and generally much more difficult to find than, *P. elegans,* perhaps because its green candles are harder to distinguish among the other plants of the scrub. The type specimen of *P. michaelii* was collected in San Luis Obispo County by G. W. Michael in June 1884.

Habitat: *Piperia michaelii* grows between sea level and 900 meters elevation, both in the coastal scrub and under nearby conifers. It also grows on grassy hillsides and in patches of grasses and poison oak in the coastal scrub. The inland colonies are often found in oak woodland.

Blooming season: *Piperia michaelii* has a fairly long blooming season. Flowering starts near the beginning of May in the interior part of its range, and lasts until the middle of August on the coast. The season depends in part on exposure to the sun: plants in full sun finish blooming nearly a month ahead of those protected by the forest, even though they may be only a few hundred meters apart. In addition to *P. elegans,* three other *Piperia* grow nearby and bloom at the same time as *P. michaelii.* Once out of the bright sun of the scrub and under the forest cover, *P. transversa* can sometimes be found in abundance, blooming within a few feet of *P. michaelii* and *P. elegans.* Deeper in the forest, *P. leptopetala* and *P. unalascensis* are sometimes blooming near inland colonies of *P. michaelii.*

Conservation: Because it occurs only rarely throughout its range, *Piperia michaelii* must be considered threatened. It is included on the Watch List, List 4, of the California Native Plant Society (Smith, Berg, et al., 1988), which means that its status should be monitored regularly for changes. Historically, *P. michaelii* is known from Los Angeles County, but the collection data sheets refer merely to hills near Glendale. That area has been heavily developed, and it is problematic that *P. michaelii* still occurs in Los Angeles County. There are also three records of *P. michaelii* in Ventura County, near the town of Ojai, but recent attempts to locate it there have been unsuccessful. This suggests that *P. michaelii* is losing some of its range, and deserves careful monitoring. Fortunately, several of its sites in the central-coast region are included in the many state and local beaches and parks along Highway 1, and some of the inland colonies are also within parks, so at least those portions of habitat are protected.

Piperia transversa Suksdorf

Allgemeine Botanische Zeitschrift Systematik 12: 43. 1906.

Etymology: The species was named in recognition of the transverse orientation of the spur in relation to the inflorescence axis.

Synonymy: none.

Common name: flat-spurred piperia.

Plate 30

Description: The flower spikes of *P. transversa* (tranz - vurs′ - ah) can reach over 30 cm in height and carry more than 90 flowers on a very slender stem. The fragrant flowers are mostly white to slightly yellowish, with green midveins on the sepals and petals. The dorsal sepal is lanceolate to ovate, the lateral sepals oblong-lanceolate. The petals are ovate-lanceolate, the lip oblong-lanceolate. *Piperia transversa* is most easily recognized by its slightly flattened horizontal spur, which measures up to 1.1 cm long, more than twice as long as the lip. The spur appears as a straight-back extension of the lip and is typically held at nearly right angles to the flower stem and ovary. Fairly dense flower spacing, coupled with the skewed spurs, creates a very full inflorescence. The dorsal sepal is usually held forward between the petals, and like the lip is aligned as if an extension of the spur. The scent, noticeable at night, is slightly reminiscent of carnations. The pair of basal leaves emerges early in spring, lies flat along the ground, and is totally withered at the time of flowering. *Piperia transversa* is pollinated by *Thallophaga taylorata* moths (see Ackerman, 1977). The ellipsoid capsules are held along the stem in a semi-erect position.

Distribution: *Piperia transversa* is found in California, Oregon, Washington, and British Columbia. Within California it occurs in 32 counties. Its range includes the mountains of Southern California in San Diego, Riverside, and San Bernardino Counties. It ranges almost continuously along the coast from San Luis Obispo County to Del Norte County, and is found down to the southern tip of the Sierras. Although *P. transversa* covers a slightly smaller range than does *P. unalascensis,* it is more plentiful where their ranges overlap.

Habitat: *Piperia transversa* grows from sea level to 2070 meters elevation. It prefers dry, open woods of mixed coniferous forest, but also grows under oaks or among shrubs. It is frequently spotted on roadbanks in very bright light, often appearing in direct sun, but occurs most often in partial shade. Along the coast, it grows just inside the shade of the forests lining the sand.

Blooming season: *Piperia transversa* blooms from late May to late August. The number of blooming plants varies from year to year in a given area. In 1986 in Yosemite National Park, several dozen plants bloomed in an area of about half an acre, but in 1987, not a single one of those plants bloomed. Peak blooming season corresponds to the peak tourist season in Yosemite, and the orchid blooms in many areas of the valley floor, mostly unnoticed by the passing thousands. In parts of its range, *P. transversa* is sympatric with *P. unalascensis, P. michaelii,* and *P. elegans. Corallorhiza maculata* and *C. striata* also bloom nearby.

Conservation: Because it is widely distributed and occurs in great numbers, *P. transversa* is not considered threatened. Large numbers of plants and significant habitat are protected in parks and reserves throughout its range.

Piperia unalascensis (Sprengel) Rydberg

Bulletin Torrey Botanical Club 28: 270. 1901.

Etymology: This species is named for Unalaska, the Aleutian Island where it was first found.

Synonymy:
Spiranthes unalascensis Sprengel, Systema Vegetabilium 3: 708. 1826.
Habenaria schischmareffiana Chamisso, Reichenbach, Linnaea 3: 29. 1828.
Herminium unalascensis (Sprengel) Orchids of Europe: 107. 1851.
Platanthera foetida Geyer ex Hooker, Kew Journal of Botany 7: 376. 1855.
Habenaria unalaschcensis (Sprengel) S. Watson, Proceedings American Academy of Arts and Sciences 12: 277. 1877.
Platanthera unalaschcensis (Sprengel) Kurtz, Botanische Jahrbücher Stst. 19: 408. 1894.
Montolivaea unalaschcensis (Sprengel) Rydberg, Memoirs New York Botanical Garden 1: 107. 1900.

Common names: Alaska piperia, slender-spire orchid.

Plate 31

Description: The stems of *P. unalascensis* (oo - na - la - sense′ - iss) can exceed 70 cm in height, sometimes carrying more than 100 tiny green flowers, but plants less than 15 cm tall can also bloom. The flowers are relatively small, only about 0.5 × 0.6 cm. The sepals vary between ovate-lanceolate and ovate-elliptic. The petals are ovate-lanceolate, the lip ovate-lanceolate to triangular-ovate. The lateral sepals curve back and clasp the spur. The spur, its position varying from horizontal to decurrent, varies in length from slightly shorter than the lip to slightly longer than the lip. A faint musky aroma—described by many as unpleasant—can be detected by getting very close to the plants. The basal leaves, usually three or four but as many as six, are angled slightly from the ground. Though the leaves are usually faded at anthesis, green leaves may remain until flowering is nearly over. Ackerman (1977) reported that *P. unalascensis* is pollinated by both pyralid moths (*Oidaematophorus* spp.) and plume moths (*Platyptilia* spp.). The ellipsoid capsules are held upright along the stem.

Distribution: *Piperia unalascensis,* growing from the mountains of Southern California to the Aleutian Islands and eastward to Quebec, has the widest range of the piperias. It also grows in the northern Lake Huron region of Ontario (Whiting and Catling, 1986). *Piperia unalascensis* is also the most widely distributed *Piperia* in California, occurring in 36 counties. It has not been reported from some of the counties where *P. transversa* grows, but occurs in several counties where *P. transversa* has not been documented. *Piperia unalascensis* is the only *Piperia* found on the eastern slope of the Sierras, although it is far more numerous on the western slope. The known coastal distribution of *P. unalascensis* was extended in 1992 when a sizable colony was discovered growing in an arroyo of the Santa Lucia Mountains in Monterey County.

Habitat: *Piperia unalascensis* has the greatest elevation range of the piperias in California, growing between 120 and 2600 meters. In portions of its range outside California, it grows even higher, reaching 3050 meters in Utah (Szczawinski, 1975). The species favors the same habitat as does *P. transversa.* It grows in open mixed or coniferous forest or under manzanita in bright to moderately shaded conditions. In some areas the plants grow in full sun, often on roadbanks and roadcuts, sometimes venturing onto the roadbed. *Piperia unalascensis* is frequently encountered while hiking, especially along trails in sequoia groves, and it is one of several orchids occasionally found in campgrounds throughout the western slopes of the Sierra Nevada. Fairly large colonies are common: one in Siskiyou County had over 60 plants within an area of 10 square meters.

Blooming season: *Piperia unalascensis* is in bloom between early May and mid-August, depending on elevation. Along the coast, it blooms before all of the other piperias except *P. cooperi*. In the Sierra Nevada, where it is most numerous, *P. unalascensis* is the last of the piperias to flower. At those higher elevations, blooming does not start until late June or early July. Throughout much of its range *P. unalascensis* is sympatric with *P. transversa*. In parts of the Sierra Nevada it blooms near *P. colemanii*, and along the coast sometimes blooms near *P. michaelii* and *P. leptopetala*.

Conservation: Because it is widely distributed throughout the state, *P. unalascensis* is not threatened at this time. Many State Parks, National Parks, and Wilderness Areas protect habitat for large numbers of plants. It is extremely rare, however, in San Bernardino and San Diego counties in Southern California.

Piperia yadonii Morgan and Ackerman

Lindleyana 5(4): 205–211. 1990.

Etymology: The species was named in honor of Vernal L. Yadon, director of the Pacific Grove Museum of Natural History.

Synonymy:
Piperia elegans (Lindley) Rydberg, Bulletin Torrey Botanical Club 28: 270. 1901 in part.

Common names: Yadon's rein-orchid, Monterey piperia.

Plate 32

Description: Flower spikes of *P. yadonii* (ya - don′ - ee - eye) reach 50 cm tall and can be loosely to densely flowered. The densely flowered plants often carry over 100 flowers. The most easily distinguished characteristic of *P. yadonii* is its bicolored upper sepal and petals. The dorsal sepal is green with white margins. The upper petals, curving toward the dorsal sepal and sometimes touching its tips, are green on the inner half, white on the outer half. The narrowly triangular lip is white and curved downward. The curved spur is about as long as the lip. The flowers have a faint sweet to musklike scent. The capsules are ellipsoid and held erect.

George Henry Grinnell, who collected many orchid specimens for herbaria, was one of the first to recognize that *P. yadonii* differed from *P. elegans*. On specimen sheets of plants he collected between 1923 and 1929,

Grinnell suggested the name *Habenaria californica,* but never published his results.

Distribution: *Piperia yadonii* is one of the rarest and most recently described orchids in California, Grinnell notwithstanding. Morgan and Ackerman (1990) described it in the same paper as *P. candida.* The species is endemic to Monterey County, with significant portions of its range extending into urban areas. The type material was collected by Abrams in 1925.

Habitat: In addition to having a very limited distribution, *P. yadonii* has a limited elevational range. All plants occur below 100 meters elevation, where it grows in pine forests and chaparral. Like most of the coastal piperias, *P. yadonii* experiences a long, dry summer. Under the pine forest, the plants receive at least partial protection from the sun, and on the edges of the forest and on roadbanks it often grows in the protection of grasses. The plants growing in the chaparral receive protection for the leaves, which fade by flowering, but the flower spikes are thrust above the vegetation and bloom in full sun.

Blooming season: The total blooming season for *P. yadonii,* is fairly short, less than two months, probably because the species is so narrowly distributed. The first flowers open in late June, and blooming is complete by early August. In parts of its range, *P. yadonii* blooms with *Epipactis helleborine* and *Spiranthes romanzoffiana.*

Conservation: Morgan and Ackerman estimate the total number of *P. yadonii* plants at about 1000. Much of the species' habitat lies within areas subject to heavy development. Because of its rarity and proximity to an urban area, *P. yadonii* deserves listing as an endangered plant, which would accord it federal and state protection. In July 1992 the U.S. Fish and Wildlife Service, Department of the Interior, took the initial steps toward adding *P. yadonii* to the Federal Endangered Species List by officially requesting data on it and five other Monterey County plants. In time *P. yadonii* may become the first wild orchid in California officially protected as an endangered plant.

11. *Platanthera* L. C. Richard

Mémoires du Muséum d'Histoire Naturelle Paris 4: 48. 1818.
Etymology: *Platanthera* derives from Greek words meaning broad or flat and anther, in reference to the wide anthers of the flowers in this genus.

Platanthera (pla - tan′ - the - ra), the most confusing orchid genus represented in California, was established by Richard in 1818, but his description was not universally followed. For example, neither Jepson (1951) nor Munz (1968) used Richard's nomenclature. But in his work on the orchids of North America, Luer (1975) recognized *Platanthera* as one of three orchid genera often included in the once cumbersome *Habenaria.* The other two are *Piperia* and *Coeloglossum.* Of the three, only *Platanthera* and *Piperia* occur in California. In the United States, *Coeloglossum* is primarily an eastern orchid, and *Habenaria,* as configured today, is found only in the extreme southeastern part of the country.

Dressler (1993) places *Platanthera* and *Habenaria* in different subtribes, Orchidinae and Habenariinae, respectively, of the tribe Orchideae, although he qualifies that placement by suggesting additional study is needed. Richard separated *Platanthera* from *Habenaria* on the basis of the broad anther of the former. *Coeloglossum* is a monotypic genus consisting only of *C. viride;* it is separated from *Platanthera* and *Habenaria* by its unique lip and nectary. Differences between *Platanthera* and *Piperia* are discussed in the treatment of *Piperia.* There are approximately 200 *Platanthera* species scattered around the world, in both hemispheres, and over 20 species grow in the United States. Leaf shape and distribution on the stem differs significantly from species to species, and even show considerable variation within *P. sparsiflora.* All the platantheras have multiple swollen or tuberoid roots that taper at both ends.

Within California we have four *Platanthera* species and at least three natural hybrids in what is often referred to as the *dilatata-hyperborea* complex. This hybrid complex has experienced a wide variety of treatments. At one extreme it was thought distinct enough to be recognized separately and was placed in the genus *Limnorchis* by Rydberg (1901). At the other extreme, all members of the *dilatata-hyperborea* complex were reduced to a variety of *P. hyperborea* (see Ames, 1910). I take an intermediate approach here by recognizing *P. dilatata* var. *leucostachys, P. hyperborea, P. stricta,* and *P. sparsiflora* as occurring in California.

There is virtually universal agreement in the literature that natural hybridizing is occurring within the *dilatata-hyperborea* complex. It is not surprising that natural hybrids should occur, given that *Platanthera* species often bloom together. The most common combinations in California are *P. dilatata* var. *leucostachys* with *P. sparsiflora, P. dilatata* var. *leucostachys* with *P. hyperborea,* and *P. dilatata* var. *leucostachys* with *P. stricta.* These growing conditions result in the three *Platanthera* hybrids known to occur in California: *P. ×estesii, P. ×lassenii,* and *P. ×media.* These hybrids and the natural variation within each species sometimes make recognizing the plants in the field or in the herbarium difficult for the novice. However, once a few identifying characteristics are understood, it is possible to reliably identify the species and varieties occurring in California.

Key to the California Species of *Platanthera*

1. Flowers white — *P. dilatata* var. *leucostachys*
1. Flowers green:
    2. Spur saccate:
        3. Lip oblong; leaves along stem — *P. stricta*
        3. Lip lanceolate; leaves clustered near bottom of stem — *P. hyperborea* var. *purpurascens*
    2. Spur not saccate:
        4. Column half or more as long as dorsal sepal; lip linear to linear-lanceolate — *P. sparsiflora*
        4. Column less than half as long as dorsal sepal; lip linear-lanceolate to ovate-lanceolate — *P. hyperborea*

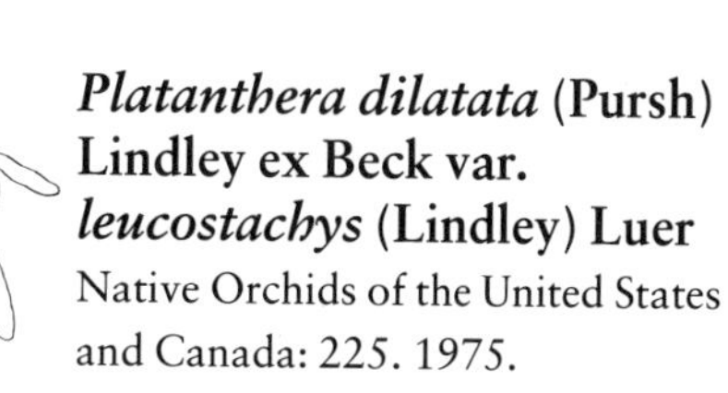

Platanthera dilatata (Pursh) Lindley ex Beck var. *leucostachys* (Lindley) Luer

Native Orchids of the United States and Canada: 225. 1975.

Etymology: The epithet *dilatata* is from the Latin word meaning broadened or expanded, in reference to the dilated base of the lip. *Leucostachys* is from Greek words meaning white spike, in reference to the white inflorescence of the variety.

Synonymy:
Platanthera leucostachys Lindley, Genera and Species of Orchidaceous Plants: 288. 1835.
Habenaria thurberi A. Gray var. *grayi* S. Watson, Proceedings American Academy of Arts and Sciences 7: 389. 1868.
Habenaria pedicellata S. Watson, Proceedings American Academy Arts and Sciences 12: 276. 1876.
Habenaria flagellans (Lindley) S. Watson in Brewer and Watson, Botany of California 2: 134. 1880.
Habenaria leucostachys (Lindley) S. Watson in Brewer and Watson, Botany of California 2: 134. 1880.
Platanthera hyperborea (Linnaeus) Lindley var. *leucostachys* (Lindley) Kranzlin, Orchidacearum Genera et Species 1: 642. 1899.
Limnorchis leucostachys (Lindley) Rydberg, Memoirs New York Botanical Garden 1: 106. 1900.
Limnorchis graminifolia Rydberg,

Bulletin Torrey Botanical Club 28: 627. 1901.
Limnorchis leucostachys (Lindley) Rydberg var. *robusta* Rydberg, Bulletin Torrey Botanical Club 28: 626. 1901.
Limnorchis thurberi (A. Gray) Rydberg, Bulletin Torrey Botanical Club 28: 624. 1901.
Habenaria dilatata (Pursh) Lindley var. *leucostachys* (Lindley) Ames, Orchidaceae 4: 71. 1910.
Limnorchis dilatata (Pursh) Rydberg var. *leucostachys* (Lindley) J. P. Anderson, Iowa State College Journal of Science 19: 189. 1945.

Common names: Sierra rein-orchid, white bog orchid, bog candles, fragrant bog orchid.

Plate 33

Description: *Platanthera dilatata* (dy - la - tah′ - ta) var. *leucostachys* (luke - oh - stak′ - ees) is the easiest to identify of the genus in California. The primary identifying characteristic is color: *P. dilatata* var. *leucostachys* is always white, and is the only white *Platanthera* in California. Other than in its color the plant is highly variable. It blooms anywhere from a mere wisp of a plant just centimeters tall to relative giants 150 cm tall with stalks nearly 2 cm thick at the base. Large plants carry over 100 flowers packed densely around the spike, and the spike of one herbarium specimen bore 248 flowers. The densely flowered raceme rising above the surrounding vegetation gives rise to some of the common names for this orchid, such as bog candles and white bog orchid. Small plants are often loosely flowered, and the specific epithet *graminifolia* has been applied to the loosely flowered form (Rydberg, 1901). The alternating leaves, up to 35 cm long, are held alongside, and partially sheath, the stem. The longest leaves are those at the base of the plant, and those near the flowers become bractlike. The flowers are very fragrant, with a hint of cloves. Individual flowers measure nearly 2 cm from the top of the dorsal sepal to the tip of the lip, and up to 2.4 cm to the apex of the spur. The dorsal sepal is ovate-elliptic, slightly concave, and connivent with the linear-lanceolate, falcate petals to form a hood over the column. The lanceolate lateral sepals are held out to the side. Most flowers exhibit slight waviness in the sepals, petals, and lip. The rhombic-lanceolate lip is widely dilated at the base, to approximately 0.3 cm wide. Lip shape, however, varies considerably. In some plants the lips widen gradually, spreading only near the base, while in others the lips dilate broadly near the middle. Lip structures between these extremes are common.

Because the petals trap the emerging lip, newly opened flowers often have a looped appearance. The lip moves into a lower position as the flower ages. The upturned lip may play a part in pollination by providing a visiting insect access to only one side of the flower, thereby ensuring that only one hemipollinarium is removed per visit (Catling and Catling, 1991). The spur varies in length but is much longer than the lip, usually from 1.5 to 2 times as long as the lip. Because of the long filiform spur, some authorities, such as Wilken and Jennings (in Hickman, ed., 1993), accept Lindley's original specific designation, and call this taxon *P. leucostachys*. The differences in the taxa, however, are so minor that Luer's varietal designation has more merit. Shorter spurs, which occur sporadically on some of our plants, have caused such plants to be mistakenly identified as *P. dilatata* (Pursh) Lindley ex Beck var. *dilatata,* which does not occur in California.

In Oregon, *P. dilatata* is pollinated by the noctuid moth *Discestra oregonica* (see Larson, 1992). The spurs of the flower carry nectar, providing a reward for the moths. While the moth drinks from the spur, its proboscis comes in contact with the viscidia and removes a pollinarium. Because the detached pollinarium is held slightly forward by the shape of the stipe, when the next flower is visited it comes in contact with the stigma, which is located above the spur entrance. Larson speculates that additional moth pollinators may be discovered via nocturnal collections. The percentage of flowers setting fruit is very high: often every flower on a stem will be pollinated. The ellipsoid capsules are held erect.

Distribution: *Platanthera dilatata* var. *leucostachys* occurs in much of the Pacific Northwest and parts of Arizona and New Mexico. Within California, it is one of our most commonly seen orchids, because of its long blooming season and its wide distribution within the state. It occurs in 40 counties, from San Diego to Del Norte, including populous Los Angeles County.

Habitat: *Platanthera dilatata* var. *leucostachys* has a wide tolerance for elevation, growing from sea-level marshes along the coast up to 3350 meters in the Sierra Nevada. It prefers full sun in wet areas such as meadows and hillside seeps, the grassy banks of a stream, or marshy areas around a lake. In all of these conditions it can also grow in partial shade, often beneath the branches of shrubs. In favorable conditions, it occasionally forms massive colonies containing thousands of plants. Because of its propensity to populate weepy roadcuts and roadside drainages, the white bog orchid is often visible to motorists.

Blooming season: The first flowers open near the beginning of May, and blooming at higher elevations can last into early September, resulting in a very long season. Unfortunately for those of us who enjoy the flowers, these plants are in the food chain of herbivores. In some areas many of the flower spikes will have been eaten off, most likely by deer, by the end of the season.

The white bog orchid blooms at the same time and in the same location as several other orchids. It can be found in meadows, by streams, and in seeps with any of the other *Platanthera*. In San Bernardino County, it blooms in the same meadow as *Malaxis monophyllos* var. *brachypoda*. On stream banks and seeps it often blooms with *Epipactis gigantea* and *Listera convallarioides*. *Spiranthes* also occasionally bloom in the same area. One of the more interesting growing companions of the white bog orchid is not another orchid, but the carnivorous *Drosera rotundifolia*. These tiny reddish plants belong to the group commonly called sundews. They have masses of sticky fingers at the ends of their leaves that attract and hold insects, which are subsequently digested by the plant. *Drosera rotundifolia* and *P. dilatata* var. *leucostachys* have overlapping ranges only from Tulare County north (Rondeau, 1991), but it is interesting to look for the sundews while orchid hunting in those areas.

Conservation: Because of its large numbers and wide distribution, *P. dilatata* var. *leucostachys* is safe from threats at this time. Large portions of its range are protected in various National Parks, State Parks, and Wilderness Areas.

Platanthera hyperborea (Linnaeus) Lindley

Genera and Species of Orchidaceous Plants: 287. 1835.

Etymology: The epithet *hyperborea* is from two Greek words meaning roughly "above the north," in reference to the region where the plant grows.

Synonymy:
Orchis hyperborea Linnaeus, Mantissa: 121. 1767.
Habenaria hyperborea (Linnaeus) R. Brown in Aiton, Hortus Kewensis, ed. 2, 5: 193. 1813.
Platanthera gracilis Lindley, Genera and Species of Orchidaceous Plants: 288. 1835.
Platanthera dilatata (Pursh) var. *gracilis* (Lindley) Ledebour, Flora Rossica 4: 71. 1853.
Limnorchis hyperborea (Linnaeus) Rydberg, Memoirs New York Botanical Garden 1: 104. 1900.
Limnorchis borealis (Chamisso) Rydberg, Bulletin Torrey Botanical Club 28: 621. 1901.
Limnorchis brachypetala Rydberg, Bulletin New York Botanical Garden 2: 161. 1901.
Limnorchis convallariaefolia (Fisher) Rydberg, Bulletin Torrey Botanical Club 28: 628. 1901.
Limnorchis gracilis (Lindley) Rydberg, Bulletin Torrey Botanical Club 28: 627. 1901.
Limnorchis laxiflora Rydberg, Bulletin Torrey Botanical Club 28: 630. 1901.

Limnorchis major (Lange) Rydberg, Bulletin Torrey Botanical Club 28: 617. 1901.
Limnorchis viridiflora (Chamisso) Rydberg, Bulletin Torrey Botanical Club 28: 616. 1901.
Habenaria viridiflora (Chamisso) Henry, Flora Southern British Columbia: 92. 1915.
Habenaria laxiflora (Rydberg) S. B. Parish, Plant World 20: 209. 1917.
Habenaria sparsiflora S. Watson var. *laxiflora* (Rydberg) Correll, Leaflets of Western Botany 3: 245. 1943.

Common name: green bog orchid.

Plate 34

Description: *Platanthera hyperborea* (hy - per - bore′ - ee - ah) is the most confusing and most diverse member of the genus growing in California. It shares many characteristics with the others of the genus and, owing to the natural variation within each species, and the intermediate hybrids, almost intergrades into them. The green bog orchid can exceed 1 meter in height and carries over 100 flowers. The lanceolate leaves, up to 20 cm long, are mostly on the lower half of the stem. The green to slightly yellowish-green flowers of *P. hyperborea* are characterized by the column, which is less than half the length of the cup formed by the dorsal sepal and petals, and a lip that tapers uniformly from the apex, or is slightly ovate, without being conspicuously dilated as is that in *P. dilatata* var. *leucostachys*. The ovaries are aligned along the inflorescence axis, and the blunt spur is variable in length but shorter than the lip. The capsules are ellipsoid and held upright along the stem. Albino forms with pure-white plant and flowers have been reported from Canada (Light and MacConaill, 1989; Szczawinski, 1975), but have not been reported in California.

Luer (1975) segregated a variety he called *P. hyperborea* var. *gracilis* (Lindley) Luer. The slender plant and distantly spaced flowers of var.

gracilis appear to be intermediate between *P. hyperborea, P. stricta,* and *P. sparsiflora.* Not all authorities agree that *P. hyperborea* var. *gracilis* is a variety of *P. hyperborea.* Some classify it as a variety of *P. stricta,* some a variety of *P. sparsiflora.* In California, *P. hyperborea* var. *gracilis* is synonymous with *P. hyperborea,* and reflects simply the normal variation in plant growth. Wherever plants corresponding to the description of *P. hyperborea* var. *gracilis* grow, it is also possible to find robust plants of normal size and flower density.

Luer (1975) also reduced a plant originally described as *Limnorchis purpurascens* Rydberg to a variety of *P. hyperborea,* calling it *P. hyperborea* var. *purpurascens* (Rydberg) Luer. The specific epithet derives from a purplish hue seen only infrequently elsewhere, but not at all in California. Our flowers are a dull yellowish green on the lip and a somewhat similar color on the inner surface of the sepals and petals, and dull green on the outer flower parts. The lip is lanceolate, as in a typical *P. hyperborea,* and the leaves are clustered near the bottom of the stem. The main difference between *P. hyperborea* var. *purpurascens* and *P. hyperborea* is in spur shape. The *P. hyperborea* var. *purpurascens* spur is much shorter than the lip and saccate. The saccate spur led Correll (1978) to include this variety within *P. stricta,* but plant habit and lip shape support its recognition as a variety of *P. hyperborea.* George Henry Grinnell discovered *P. hyperborea* var. *purpurascens* in California in July 1923.

As with several of our other wild orchids, there are few data on the pollination of *P. hyperborea.* Crandall (1900), however, reported observing mosquitoes pollinating *P. hyperborea* in Colorado.

Distribution: *Platanthera hyperborea* is by far the most widely distributed member of the genus outside California, growing in much of Canada and Alaska and parts of Asia. In the United States it grows in most of the Western states and from the Great Lakes region to New England. Within California, its distribution is limited to only seven counties. It grows in the northern counties of Del Norte, Humboldt, Siskiyou, and Trinity, and in the eastern Sierras and White Mountains in Inyo, Mono, and Tulare counties. Within California, *P. hyperborea* var. *purpurascens* grows only in Mariposa and Modoc Counties; its next-nearest location is in eastern Arizona.

Habitat: *Platanthera hyperborea* grows only at elevations between

1300 and 3200 meters, but has a slightly wider tolerance for habitat than do the other platantheras in California. Its typical habitat is grassy, damp stream banks and damp meadows, most often in nearly full sun, but occasionally in partial shade. It also grows in dry conditions on hill-sides under shrubs and pines. Quite possibly, these dry areas are wet in early spring, but have completely dried out by the late summer bloom of *P. hyperborea*. *Platanthera hyperborea* var. *purpurascens* grows in wet meadows at an elevation of 2100 meters.

Blooming season: Depending on elevation, *P. hyperborea* can be found in bloom from early June to the latter part of August. Throughout much of its limited range in California, it shares habitat with *P. dilatata* var. *leuc-ostachys*, and also blooms with *E. gigantea*. Like those of *P. dilatata* var. *leucostachys*, the flower spikes of *P. hyperborea* are often eaten by deer, which can be very frustrating when you have driven 500 km to see the flowers. *Platanthera hyperborea* var. *purpurascens* blooms from the be-ginning of July to early August.

Conservation: *Platanthera hyperborea* is not common in California, but it exists in sufficient numbers and in such diverse habitat that it is not threatened, especially in the White Mountains and eastern Sierras. In the northern counties, however, it is far less numerous, and is reported only rarely. It occurs in heavily logged areas in that part of the state, and deserves continued study and frequent review of its status. *Platanthera hyperborea* var. *purpurascens* is extremely rare and surely endangered in California. The only known colony in Mariposa County, consisting of less than ten plants, grows in a meadow frequented by tourists, and it is there-fore subject to accidental loss by trampling.

Notes and comments: *Platanthera hyperborea* var. *purpurascens* was not included in California's flora until 1993. Leon Glicenstein, a frequent orchid-hunting companion, and I came across two specimen sheets of it while studying some unidentified *Platanthera* collections at the Rancho Santa Ana Botanic Garden's herbarium. George Henry Grinnell had col-lected the specimens on the same day in 1923, but he had not been able to identify the plants. His notes imply that both sheets were from the same location, that one sheet was for his private herbarium, and that the other was for the University of Southern California's herbarium. Both Grinnell's collection and that of the University of Southern California were subsequently transferred to the herbarium of the Rancho Santa Ana Botanic Garden. In July 1993, an extensive search of Grinnell's collection area resulted in the discovery of one small colony of the plants, the only known colony in the county. Charles Sheviak of the New York State Museum confirmed our tentative identification of the plants. Glicenstein has also found *P. hyperborea* var. *purpurascens* growing in Modoc County.

Platanthera sparsiflora (S. Watson) Schlechter

Bulletin l'Herbier Boissier 7: 538. 1899.

Etymology: The epithet *sparsiflora*, from Latin, means "scattered flowers" or sparsely flowered, in reference to the supposedly laxly flowered spikes.

Synonymy:
Habenaria sparsiflora S. Watson, Proceedings American Academy of Arts and Sciences 11: 276. 1877.
Limnorchis sparsiflora (S. Watson) Rydberg, Bulletin Torrey Botanical Club 28: 631. 1901.
Habenaria aggregata Howell, Flora of Northwest America: 628. 1902.
Limnorchis aggregata (Howell) Frye and Rigg, Northwest Flora: 114. 1912.
Habenaria leucostachys (Lindley) S. Watson var. *viridis* Jepson, Flora of California: 331. 1921.

Common name: sparsely flowered bog orchid.

Plate 35

Description: The typical *P. sparsiflora* (spar - si - flor′ - a) is from 40 to 60 cm tall, but is occasionally as tall as 94 cm, or blooms at as little as 8 cm. Its common name is the sparsely flowered bog orchid, but the sparseness must be interpreted in terms of *P. dilatata* var. *leucostachys,* because *P. sparsiflora* sometimes bears over 120 green to very yellowish green flowers per plant. *Platanthera sparsiflora* is most easily recognized by its large column, which fills half or more of the hood formed by the ovate-elliptic dorsal sepal and the ovate-lanceolate, falcate petals. Usually, the linear-lanceolate lower sepals are reflexed back, and sometimes are twisted, giving the flowers a narrow appearance. The lip, which varies in length, is linear though variable in width, and thickened at the base. (Flowers that would otherwise be identified as *P. sparsiflora* but have lips that are not linear, or have a smaller than normal column, may be the result of gene flow from *P. dilatata* var. *leucostachys*.) The spur is variable in length, from slightly shorter than the lip to much longer, up to 1.5 times as long as the lip. The flowers have a strong aroma. The ovaries angle out from the inflorescence axis, resulting in a slight downward tilt to the flowers. The ellipsoid capsules remain in that semi-erect position.

Plant structure in *P. sparsiflora* is variable. Typical stems support evenly spaced lanceolate leaves, but throughout the species' range in California some plants have longer lanceolate leaves clustered near the bottom of the stem, in the manner of *P. dilatata* var. *leucostachys* or *P. hyperborea*. Luer (1975) called such plants *P. sparsiflora* var. *ensifolia* (Rydberg) Luer.

Correll (1978) identified a specimen of *P. sparsiflora* var. *laxiflora* (Rydberg) Correll as occurring in California. The plant he describes sounds very much like *P. stricta* with a long, slender spur. However, after studying the specimens Correll annotated at the California Academy of Science, I conclude it is clear that *P. sparsiflora* var. *laxiflora* is the same plant Luer calls *P. hyperborea* var. *gracilis*.

The pollination biology of *P. sparsiflora* has been incompletely studied. Kipping (1971) reported that it is pollinated by moths, the pollinaria attaching to the moth's proboscis. Fruit set is high, approaching 100% in most of the range.

Distribution: *Platanthera sparsiflora*, the most common green *Platanthera* in California, is distributed from San Diego County to Del Norte County. It covers most of the same range as *P. dilatata* var. *leucostachys*, but has been documented in just 31 counties, nine fewer. Watson (1876) described *P. sparsiflora* from material collected in the Sierra Nevada of Northern California. *Platanthera sparsiflora* grows in most of the western states, from Washington to New Mexico, and in much of Mexico.

Habitat: *Platanthera sparsiflora* grows between 100 and 3350 meters elevation, and is thus one of our highest-elevation orchids. The sparsely flowered bog orchid requires wet habitat such as damp meadows, stream banks, wet hillsides, and roadside ditches and seeps. It usually grows in full sun, but is sometimes found in the partial shade of bushes or trees lining streams or bordering meadows. Because its plants and flowers are hard to distinguish among the grasses with which it grows, *P. sparsiflora* can be difficult to see, even where it occurs in great numbers.

Blooming season: The blooming season for the sparsely flowered bog orchid stretches from the middle of May to the first of September. Because it blooms during peak tourist season, summer vacationers can always expect to find *P. sparsiflora* in flower. Early in the season, the sparsely flowered bog orchid blooms along with *Cypripedium californicum,* and late in the orchid year it can still be blooming in the company of *Listera convallarioides.* Often in bloom in the same habitat are *P. dilatata* var. *leucostachys* and *Corallorhiza trifida* var. *verna. Corallorhiza maculata* and *C. striata* bloom in drier areas nearby, as does *Goodyera oblongifolia,* though later in the season.

Conservation: The sparsely flowered bog orchid exists in sufficient numbers and sufficiently scattered locations to be safe from immediate threats. Significant portions of its habitat are protected within State Parks and National Parks.

Platanthera stricta Lindley

Genera and Species of Orchidaceous Plants: 288. 1835.

Etymology: The epithet *stricta* is from the Latin for straight or narrow, in reference to the habit of the spike.

Synonymy:
Platanthera dilatata (Pursh) Lindley var. *gracilis* Ledebour, Flora Rossica 4: 71. 1853.
Habenaria gracilis (Ledebour) S. Watson, Proceedings American Academy of Arts and Sciences 11: 277. 1876.
Habenaria saccata Greene, Erythea 3: 49. 1895.
Habenaria stricta (Lindley) Rydberg, Bulletin Torrey Botanical Club 24: 189. 1897.
Limnorchis stricta (Lindley) Rydberg, Memoirs New York Botanical Garden 1: 105. 1900.
Platanthera saccata (Greene) Hultén, Madroño 19: 223. 1968.

Common name: slender bog orchid.

Plate 36

Description: *Platanthera stricta* (strik′ - ta) is fairly easy to recognize. The flowers on these plants can be unambiguously identified by their saccate spur. The shape of the spur is the source of the name *P. saccata* (Greene) Hultén, which is often used for this species. The plants, 60 to 90 cm tall, can bear upwards of 60 flowers, and the flowers are usually green, though the spurs occasionally show a purplish tinge. The lip is narrowly oblong to somewhat elliptic. The lateral sepals and petals are elliptic-lanceolate, and the lateral sepals are often twisted and slightly reflexed.

The column is much smaller than the hood formed by the petals and ovate-elliptic dorsal sepal. Typical plants have shorter leaves than the other California members of the genus, but generally resemble *P. sparsiflora* in structure. The leaves, 6 to 8 cm long, are evenly spaced and held horizontally. Plant structure and lip shape help differentiate *P. stricta* from *P. hyperborea* var. *purpurascens,* the only other saccate-spurred *Platanthera* in California. The ellipsoid capsules are held erect.

Spur length is variable in *P. stricta.* In some plants the spur starts out slender, much like a spur on *P. hyperborea,* but then terminates in a bulbous sack. The length of these elongated spurs approaches the length of the lip. On other plants there is little or no cylindrical portion to the spur; it consists entirely of a sack about half the length of the lip or shorter. Plants with lips dilated at the base may be hybrids with *P. hyperborea* or *P. dilatata* var. *leucostachys.*

Platanthera stricta is pollinated by a variety of insects, including flies of the genera *Empis* and *Rhamphomyia.* Other pollinators include the fly *Anthepiscopus longipalpis,* the bees *Bombus melanopygus* and *B. flavifrons,* and the moth *Eustroma fasciata* (see Patt et al., 1989). All of these insects have relatively short mouth parts, and typically forage on the inflorescence for droplets of nectar before probing the spur. The foraging and probing bring the insects into frequent contact with the column, effecting the removal and deposition of pollinaria. Fifty-two percent of the flowers in Patt's study group set seed capsules.

Distribution: *Platanthera stricta* grows only in the northwestern United States, along the west coast of Canada, and on into Alaska. Within California, *P. stricta* occurs only in five counties: Del Norte, Humboldt, Siskiyou, Trinity, and Modoc. Greene (1895) described *Habenaria saccata* on the basis of material collected by Mrs. R. M. Austin from Lassen Creek in Modoc County.

Habitat: Unlike the other platantheras in California, *P. stricta* occupies a rather narrow elevation band, growing only between 1000 and 2300 meters. This pattern may reflect simply its limited occurrence in our state, because in other areas of its range it grows at higher elevations, reaching 3050 meters in Colorado (see Long, 1970). *Platanthera stricta* favors the same wet habitat as the other platantheras. Typically, it grows in moist meadows, seeps, and along streams, and it thrives equally well in full sun or partly shaded areas. In Modoc County it blooms along hot mineral springs and near bubbling mudpots. *Platanthera stricta* can be locally plentiful but nonetheless hard to find, because it blends in so well with the grasses that share its habitats.

Blooming season: *Platanthera stricta* has a long blooming season. It starts blooming in mid-May and at higher elevations can still be found in bloom in early August. Because individual flowers last for about three weeks, and a typical inflorescence takes about three weeks to open, a single plant can remain in bloom for up to six weeks. *Platanthera stricta* blooms at the same time and in the same habitat as *P. dilatata* var. *leucostachys*, and it is common to find the two blooming side by side. *Spiranthes romanzoffiana* and *Piperia unalascensis* also bloom nearby.

Conservation: *Platanthera stricta* should be considered threatened in California because it occurs in such a small area. Except where it occurs in Modoc County it is rarely seen. A few colonies are within Wilderness Areas, but most of the plants grow on National Forest lands subject to logging.

Platanthera ×*estesii* Schrenk

Die Orchidee 26: 261. 1975.

Plate 39

The shared habitat and common blooming season of *P. stricta* and *P. dilatata* var. *leucostachys* result in the hybrid *P.* ×*estesii* (*P. dilatata* × *P. stricta*). The most common hybrid flowers are a whitish green and resemble a *P. stricta* with a dilated lip and a longer than normal saccate spur. Other structural anomalies may occur, including misshapen, incompletely developed flowers. On some flowers the spur resembles that of the *P. dilatata* var. *leucostachys* parent. *Platanthera* ×*estesii* grows in the South Warner Wilderness of Modoc County.

Platanthera ×*lassenii* Schrenk

Die Orchidee 26: 261. 1975.

Plate 39

Over much of their range, *P. dilatata* var. *leucostachys* and *P. sparsiflora* overlap in both distribution and blooming season. Occasionally, a natural hybrid between the two occurs. *Platanthera* ×*lassenii* (*P. dilatata* × *P. sparsiflora*) exhibits characteristics of both parents. Spur length is variable, but the spur is usually longer than the lip. The column is intermediate in size between those of the two parents. The lip usually shows some of the dilatation of its *P. dilatata* heritage. The sepals and petals are green, but a paler green than is usual for *P. sparsiflora,* and the spur and lip are a whitish green. Because of its green color and dilated lip, *P.* ×*lassenii* is probably responsible for the literature reports of green forms of *P. dilatata.* The hybrid is commonly found growing with both parents. Sometimes, plants with misshapen flowers result from the cross, and grow in hybrid swarms with the parents and well-formed hybrids. The type specimen for *P.* ×*lassenii* is from Lassen Volcanic National Park, and the plants also grow in Mono, Fresno, and Los Angeles Counties, and can be expected wherever the two species occur together.

Platanthera ×*media* (Rydberg) Luer

Native Orchids of the United States and Canada 229. 1975.

Plate 39

The most common *Platanthera* hybrid in most of the United States, *P.* ×*media* (Rydberg) Luer, which is *P. dilatata* × *P. hyperborea,* occurs only rarely in California. It is known only from a 1993 discovery in Mono County by Leon Glicenstein. This natural hybrid can be difficult to identify, because except for color and the widely dilated lip of *P. dilatata* var. *leucostachys,* the two parents have many characteristics in common. The pale whitish-green of the hybrid is readily recognizable, however, and close study will reveal that the lip is more dilated than on most *P. hyperborea,* and not as dilated as on most *P. dilatata* var. *leucostachys. Platanthera* ×*media* can be expected wherever the ranges of the parents overlap.

12. *Spiranthes* L. C. Richard

Mémoires du Muséum d'Histoire Naturelle Paris 4: 50. 1818.
Etymology: *Spiranthes* is from two Greek words meaning coil and flowers, in reference to the coiled or spiraled flower spikes of the genus.

Spiranthes (spy - ran′ - theez) is a member of a large and confusing complex. In his revision of the subfamily Spiranthinae, Garay (1980) identified 390 species in 44 genera, and recognized 42 species of *Spiranthes* scattered throughout much of the Northern and Southern Hemispheres. About 20 species occur in the United States and Canada. In an apparent allusion to the resemblance of the floral spirals to certain hair styles, spiranthes are commonly called ladies' tresses.

Both of the species that occur in California belong to the *Spiranthes cernua* complex, which has undergone revision lately, especially by Sheviak (1982). The history of *Spiranthes* includes alternating periods of splitting and lumping, and the California ladies' tresses are a good example of the various treatments. Ames and Correll (1943) considered *S. porrifolia* to be only a variety of *S. romanzoffiana,* calling it *S. romanzoffiana* var. *porrifolia* (Lindley) Ames and Correll. They cited intergrading of lip shape and calli characteristics as reasons for combining the two species. Field identification of the species is sometimes difficult because of an intermediate and yet unnamed form that grows in the Sierra Nevada in Mariposa and Tuolomne Counties and perhaps elsewhere, but the differences between the taxa are sufficient for us to continue to recognize both *S. porrifolia* and *S. romanzoffiana.* Additional studies are required to determine if the intermediate form in the Sierras represents a hybrid, a variety, or a third California species.

The pollination mechanics of *Spiranthes* were first described by Darwin (1877). The procedures and the protection against self-pollination are much the same as in *Goodyera*. On freshly opened flowers, the column is positioned close to the lip below, blocking entrance to the nectary. A visiting insect, probing for nectar, instead comes in contact with the pollinia and bears some away on its head or back. As the flower ages, the relative separation of the column and lip increases, allowing pollinators to reach the nectary. Any pollinia on an insect at that point come in contact with the sticky stigma and are effectively stripped off. Pollinators begin their visits at the bottom of the flower spike, first contacting the mature flowers and depositing there whatever pollinia they may be carrying. As they work their way up the spike, they encounter newly opened flowers, and pick up a new load of pollinia for the next spike they visit.

Three new *Spiranthes* from the western United States have been described in recent years (see Sheviak, 1984, 1989, 1990b). One of these, *Spiranthes infernalis* Sheviak, is from Nye County, Nevada, just over the border from the Death Valley region of California, and thus may occur in California as well. Sheviak also suggests that one additional species may be included in the California taxa because of differences in the coastal plants currently included within *S. romanzoffiana*.

Key to the California Species of *Spiranthes*

1. Lip pandurate, the apex dilated; sepals and petals united throughout their lengths and forming a hood ... *S. romanzoffiana*
1. Lip ovate to lanceolate, the apex only slightly or not at all dilated; apices of sepals and petals free and spreading ... *S. porrifolia*

Spiranthes porrifolia Lindley

Genera and Species of Orchidaceous Plants: 467. 1840.

Etymology: The epithet *porrifolia* is from two Latin words meaning leek green and leaves, in reference to the color of the leaves.

Synonymy:
Gyrostachys porrifolia (Lindley) Kuntze, Reviso Generum Plantarum 2: 664. 1891.
Orchiastrum porrifolium (Lindley) Greene, Manual of the Botany of the Region of San Francisco Bay: 306. 1894.
Ibidium porrifolium Rydberg, Bulletin Torrey Botanical Club 32: 610. 1905.
Spiranthes romanzoffiana Chamisso var. *porrifolia* (Lindley) Ames and Correll, Botanical Museum Leaflets, Harvard University 11: 1. 1943.

Common name: western ladies' tresses.

Plate 37

Description: *Spiranthes porrifolia* (pore - i - foe′ - lee - ah) varies in size from under 10 to over 74 cm tall. The three to five basal leaves rise from swollen tuberoid roots, and are as much as 25 cm long. On the tallest plants the upper leaves spread along the lower portion of the stem, but on smaller plants the leaves are all clustered at the base. The racemes are sometimes loosely wound, but often have tight, multiple spirals of over 100 creamy-yellow flowers on a single stem. The ovate lip varies from plant to plant: on some it tapers gently to a point; on others it spreads slightly at the tip. The sepals and petals, all relatively lanceolate, are free and spreading at the apices. *Spiranthes porrifolia* is best recognized by the prominent tuberosities at the base of the lip, near the junction with the ovary, but the flower must be dissected before they can be seen. The ellipsoid capsules are held semi-erect.

Distribution: Of our two ladies' tresses, *S. porrifolia* is the more restricted in range. It is limited to the western regions of the United States, primarily California, Oregon, and Washington, which is how it earned the common name of western ladies' tresses. Within California, it is very widespread, occurring in 35 counties. *Spiranthes porrifolia* is most easily found north of Monterey and Tulare Counties, and is extremely rare in Southern California.

Habitat: *Spiranthes porrifolia* grows in two different major habitats, from near sea level to over 2500 meters elevation. A constant source of water is a common characteristic of the two. Perhaps the more usual habitat is in full sun in damp meadows or along streams. The grasses provide a protecting cover, and *S. porrifolia* can be virtually impossible to find when it is not in bloom. Its other major habitat is in wet spots on steep, rocky cliffs. Usually, these wet spots are easy to pick out because of the green growth associated with the water in an otherwise dry landscape. At other times, the damp soil stands out as darker than the surrounding dry walls. The plants root in the dirt between rocks, and are very difficult to detect without close inspection. The species also grows on alkali flats at the northern end of Owens Valley.

Blooming season: *Spiranthes porrifolia* blooms between the end of May and the beginning of September, and the peak of its season is mid-June to mid-July. The ranges of *S. porrifolia* and *S. romanzoffiana* overlap significantly, but in the areas they share, *S. porrifolia* typically has an earlier blooming season, though the seasons do coincide somewhat. Their overlapping blooming seasons support the argument that intermediate forms between *S. porrifolia* and *S. romanzoffiana* may be hybrids. The near-vertical cliffs favored by *S. porrifolia* are also home to *Epipactis gigantea,* and on some damp cliff walls they bloom together.

Conservation: Because of its widespread distribution and its ability to establish itself on steep cliffs, *S. porrifolia* is relatively safe from threats. Portions of its range are conserved within State Parks and National Parks, but it does appear to be in danger in part of its range. Herbarium collections document *S. porrifolia* in the mountains of San Diego and San Bernardino Counties in Southern California. Fultz (1928) even reported it as common in areas of the San Bernardino Mountains. Recent attempts to locate the plant in Southern California have been unsuccessful, however, and it is possible that *S. porrifolia* has lost some of its range.

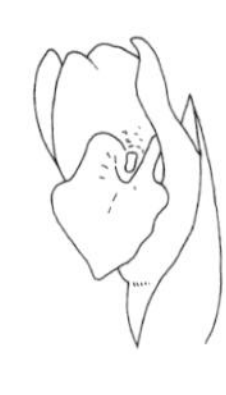

Spiranthes romanzoffiana Chamisso

Linnaea 3: 32. 1828.

Etymology: The specific epithet honors Nicholas Romanzof, a Russian minister of state. That Romanzof would be honored in this manner reflects the fact that this species was discovered in Alaska at a time when Alaska was a Russian territory.

Synonymy:
Gyrostachys romanzowiana (Chamisso) MacMillan, Metasp. Minn. Valley: 171. 1892.
Orchiastrum romanzoffianum (Chamisso) Greene, Manual of the Botany of the Region of San Francisco Bay: 306. 1894.
Gyrostachys stricta Rydberg, Memoirs New York Botanical Garden 1: 107. 1900.
Ibidium strictum (Rydberg) House, Bulletin Torrey Botanical Club 32: 381. 1905.
Ibidium romanzoffianum (Chamisso) House, Muhlenbergia 1: 129. 1906.
Spiranthes stricta (Rydberg) A. Nelson, New Manual of Botany, Rocky Mountains: 125. 1909.
Triorchis romanzoffiana (Chamisso) Nieuwland, American Midland Naturalist 3: 123. 1913.
Triorchis stricta (Rydberg) Nieuwland, American Midland Naturalist 3: 123. 1913.

Common name: hooded ladies' tresses.

Plate 38

Description: Blooming plants of *S. romanzoffiana* (ro - man - zoff - ee - ah′ - na) vary in height with habitat and elevation. The tallest inflorescences rarely exceed 46 cm, although Larson and Larson (1987) reported inflorescences up to 57 cm tall. At its elevation limit, well above timberline, there is a short growing season and deep snow cover for much of the year. As a result, the plants at higher elevations are small, blooming at under 6 cm. The three to six leaves of the hooded ladies' tresses are mostly basal, although there are multiple bracts along the stem. The leaves sprout from a pair of tuberoid roots. The classic *S. romanzoffiana* has a tightly wound inflorescence of up to 60 pristine white flowers arranged in three spiraling rows. Three features distinguish the species from the other spiranthes. The lanceolate sepals and linear petals form a tight hood over the downward-curving lip, giving rise to the common name hooded ladies' tresses. Lip shape, usually referred to as pandurate, or fiddle-shaped, is the second character. The lip narrows noticeably about two-thirds the distance from the base before spreading out at the tip, although, as with *S. porrifolia,* this is not a definitive characteristic: on some plants the lip is as wide as the full flower; on others the lip is indistinguishable from the slightly spreading lip of *S. porrifolia.* To see the third identifying characteristic of the species, it is necessary to dissect the flower: at the base of the lip *S. romanzoffiana* has minute tuberosities resembling slight swellings. The ellipsoid capsules are held semi-erect.

Plants along the coast tend to have larger flowers than do the inland plants, and are often leafless in bloom. Sheviak (1990b) provided a separate key for these coastal plants, speculating that they may eventually prove to be a new taxon.

Spiranthes romanzoffiana gives off a sweet aroma and attracts plenty of pollinators. Bingham (1939) describes the fragrance as being "of delicate lilacs." *Spiranthes romanzoffiana* is not pollinator-specific. Larson and Larson (1987, 1990) identified 11 different pollinators: six species of *Bombus* (bumblebees) including, *B. bifarius,* one cuckoo bee, one leaf-cutting bee, and three halictid bees. The bees visited multiple times over long periods, successfully pollinating over 75% of the flowers. Visiting bumblebees land on the lowest flowers first and work their way up the inflorescence. The tallest and showiest plants attracted the most visitors. Six weeks after pollination the capsules had dehisced and were dispersing seed.

Distribution: *Spiranthes romanzoffiana* is much more widely distributed in the United States than is *S. porrifolia.* Its range overlaps that of *S. porrifolia* in the west and extends across Canada to the Atlantic. It also occurs in Ireland, where Godfery (1922) reported it exceedingly rare and local. Within California it occurs in 36 counties. Inland it ranges as far south as Riverside County in Southern California, and its southern terminus along the coast is in the northern part of Santa Barbara County. Abrams (1904), Parsons (1907), and Fultz (1928) all reported *S. romanzoffiana* from the canyons and foothills of Los Angeles County, but the herbarium collection records do not support their claims, and the possibility that the hooded ladies' tresses persists in Los Angeles County is therefore discounted.

Habitat: *Spiranthes romanzoffiana* thrives in several different habitats, mostly in full sun, between sea level and 3260 meters elevation. In the mountains, its most common habitat is meadows. Early in the spring and summer, the meadows are wet, boggy areas, and it is when the meadows begin to dry out, at the end of summer, that *S. romanzoffiana* starts to bloom. It can first be found at the very edges of the meadows, near tree line, and as fall approaches, plants toward the center of the meadow begin to bloom. A second habitat for the species is along the banks of streams, and on some of our mountain lakes it grows on floating sphagnum bogs.

The hooded ladies' tresses also grows in multiple coastal habitats. Its most unusual coastal habitat is from Santa Barbara County northward, where *S. romanzoffiana* blooms in dry, sun-baked coastal grassland. Perhaps the areas where it grows are wet enough during the rainy season, but by the time the orchids bloom the relentless sun has baked the soil dry. *Spiranthes romanzoffiana* also manages to survive on the tops of cliffs just above the ocean, and on the walls of the cliffs leading down to coastal lagoons. Plants in this habitat tend to have the largest flowers and the densest inflorescences, and are often leafless in bloom. These variable characteristics are the reason Sheviak (1990b) provided a separate key for them. If *S. romanzoffiana* did at one time occur in the foothills of Los Angeles County, it was probably in this dry, coastal grassland habitat. Grassy swales in sand dunes are another of the species' seacoast habitats' and it can also grow in the dry woods of coastal pine forests where there is at least partial shade.

Blooming season: *Spiranthes romanzoffiana* is the one of the last orchids to bloom in California. It starts blooming in mid-June, and in portions of its range it will still be blooming in late September. In the Sierra Nevada, it is a reliable late-season bloomer. Several other orchids share the same habitat as *S. romanzoffiana.* The early orchids such as *Platanthera dilatata* var. *leucostachys* and *P. sparsiflora* thrive in the same meadows, and bloom at the same time. In its dry coastal habitat, *S. romanzoffiana* blooms with *Piperia elegans.*

Conservation: Because of its wide distribution and high-elevation growth pattern, *S. romanzoffiana* is safe from threats, when considered across its entire range, but examples of losses are many. One colony in San Luis Obispo County was destroyed by the dumping of material from road repairs, and the largest colony in Santa Cruz Colony is located on land scheduled for development. The hooded ladies' tresses is extremely rare in San Bernardino and San Diego Counties, and warrants frequent monitoring of its status.

Notes and Comments: The disjunct location of *S. romanzoffiana* in Ireland, where it is called Irish ladies' tresses, deserves some comment. Both Summerhayes (1968) and Davies et al. (1988) speculate that its occurrence there is a relic of older, colder times, when *S. romanzoffiana* must have been more widespread. Davies notes that it occurs in Ireland in association with two other North American plants, and suggests that the three plants may be what remains from a more extensive flora predating the last ice age. The occurrence of *S. romanzoffiana* in Ireland is not unlike the disjunct occurrences in California of *Malaxis monophyllos* var. *brachypoda* in San Bernardino County and of *C. trifida* in Plumas County. They too are surviving in a habitat niche far removed from their main areas of distribution, which may indicate that they were once more common.

Spiranthes undescribed

Plate 39

Scattered through much of the Sierra Nevada, especially in Mariposa, Madera, and Tuolumne Counties, is a *Spiranthes* that does not fit the descriptions of the two species that occur in California. The lip, sepals, and petals are all intermediate between those of *S. porrifolia* and *S. romanzoffiana.* The lateral sepals are free and spreading for most of their length, but are connate for about one-third of their length where they meet under the lip. The petals are united and form, with the dorsal sepal, a tight hood. The lip is either pointed or only slightly spread at the tip. The plants grow in meadows, along streams, and in roadside ditches.

One postulate is that this plant is a hybrid between *S. porrifolia* and *S. romanzoffiana,* but beause it occurs in massive colonies in the absence of either of the other species, it appears to be independently established and self-propogating. Clearly, additional study is required if we are to determine whether these plants constitute a hybrid, a variety, or even a separate species.

Appendix 1. Herbarium Documentation of Distribution of California's Wild Orchids, by County, Including Channel Islands

The indicated occurrence of a species in a county is in almost every case documented by a pressed specimen in one of the many herbaria in California or elsewhere. Although more than one specimen exists for most species in most counties, only one per county is indicated in the table. In some cases, indicated by "RAC" (Ronald A. Coleman), no herbarium specimens are known to exist for documentation in those counties, but my field searches have verified that the plants in question grow in the counties in question.

The following institutions (listed alphabetically by abbreviation used in the table) permitted me to study their herbarium orchid collections: CAS = California Academy of Science, San Francisco; CHSC = California State University, Chico; CPH = University of the Pacific; DS = Dudley Herbarium, Stanford University (now at CAS); HSC = California State University, Humboldt; JEP = Jepson Herbarium, now at University of California, Berkeley; LA = University of California, Los Angeles; MACF = California State University, Fullerton; NYBG = New York Botanic Garden; OBI = California State Polytechnic University, San Luis Obispo; PGMNH = Pacific Grove Museum of Natural History; PUA = Pacific Union College; ROPA = California State University, Sonoma; RSA = Rancho Santa Ana Botanic Garden; SACT = California State University, Sacramento; SBBG = Santa Barbara Botanic Garden; SD = San Diego Museum of Natural History; SFSU = California State University, San Francisco; SFV = California State University, Northridge; UC = University of California, Berkeley; UCR = University of California, Riverside; UCSB = University of California, Santa Barbara; US = United States National Herbarium, Washington.

Species	Alameda	Alpine	Amador	Butte	Calaveras	Channel Islands
Calypso bulbosa						
Cephalanthera austiniae				RSA	JEP/UC	
Corallorhiza maculata	RSA		RSA	RSA	RSA	
Corallorhiza mertensiana				CSU		
Corallorhiza striata			CAS	US	CAS	
Corallorhiza trifida						
Cypripedium californicum						
Cypripedium fasciculatum				CAS		
Cypripedium montanum						
Epipactis gigantea		RSA		CAS	RSA	RSA
Epipactis helleborine	CAS					
Goodyera oblongifolia				RSA	CAS	
Listera caurina						
Listera convallarioides		US		RSA	CAS	
Listera cordata						
Malaxis monophyllos						
Piperia candida						
Piperia colemanii				JEP/UC		
Piperia cooperi						CAS
Piperia elegans	CAS					
Piperia elongata	JEP/UC		RSA	CAS	JEP/UC	RSA
Piperia leptopetala						
Piperia michaelii	JEP/UC		CAS	CSU		
Piperia transversa			CAS	CAS	CAS	
Piperia unalascensis		CAS	CAS	CAS	CAS	
Piperia yadonii						
Platanthera dilatata		RSA	CAS	RSA	CAS	
Platanthera hyperborea						
Platanthera sparsiflora		JEP/UC	UOP	JEP/UC	CAS	
Platanthera stricta						
Spiranthes porrifolia	JEP/UC	SBBG	CAS	CAS	RSA	
Spiranthes romanzoffiana		CAS		CAS	JEP/UC	

Species	Colusa	Contra Costa	Del Norte	El Dorado	Fresno	Glenn
Calypso bulbosa			RSA			
Cephalanthera austiniae	JEP/UC		RSA	JEP/UC	JEP/UC	CAS
Corallorhiza maculata		UOP	CAS	RSA	RSA	CAS
Corallorhiza mertensiana			RSA	CAS		
Corallorhiza striata				CAS	RSA	CAS
Corallorhiza trifida						
Cypripedium californicum			RSA			
Cypripedium fasciculatum			RSA			
Cypripedium montanum			CAS			
Epipactis gigantea	JEP/UC	CAS	RSA	JEP/UC	RSA	JEP/UC
Epipactis helleborine		JEP/UC				
Goodyera oblongifolia			RSA	CAS	RSA	
Listera caurina			RSA			
Listera convallarioides	JEP/UC		RSA	RSA	RSA	CAS
Listera cordata			RSA			
Malaxis monophyllos						
Piperia candida			RSA			
Piperia colemanii	JEP/UC				RSA	
Piperia cooperi						
Piperia elegans		CAS	CAS			
Piperia elongata		CAS	RSA	UCSB		
Piperia leptopetala						
Piperia michaelii		JEP/UC			JEP/UC	
Piperia transversa			CAS	CAS	CAS	
Piperia unalascensis		JEP/UC	CAS	JEP/UC	RSA	CAS
Piperia yadonii						
Platanthera dilatata			CAS	RSA	RSA	RSA
Platanthera hyperborea			HSU			
Platanthera sparsiflora			RSA	RSA	RSA	CAS
Platanthera stricta			CAS			
Spiranthes porrifolia			RSA	JEP/UC	CAS	
Spiranthes romanzoffiana		JEP/UC	RSA	RSA	RSA	CAS

Species	Humboldt	Imperial	Inyo	Kern	Kings	Lake
Calypso bulbosa	CAS					SFSU
Cephalanthera austiniae	RSA					CAS
Corallorhiza maculata	RSA		RSA	RSA		CAS
Corallorhiza mertensiana	CAS					
Corallorhiza striata	RSA					RSA
Corallorhiza trifida						
Cypripedium californicum	RSA					
Cypripedium fasciculatum	JEP/UC					
Cypripedium montanum	CAS					
Epipactis gigantea	RSA	UCR	UCSB	CAS		CAS
Epipactis helleborine						
Goodyera oblongifolia	RSA					CAS
Listera caurina	RSA					
Listera convallarioides	JEP/UC		CAS	RSA		RSA
Listera cordata	JEP/UC					
Malaxis monophyllos						
Piperia candida	CAS					
Piperia colemanii						
Piperia cooperi						
Piperia elegans	RSA					CAS
Piperia elongata	RSA					RSA
Piperia leptopetala						CAS
Piperia michaelii	JEP/UC					
Piperia transversa	CAS					RSA
Piperia unalascensis	CAS			RSA		CAS
Piperia yadonii						
Platanthera dilatata	RSA		CAS	RSA		RSA
Platanthera hyperborea	ROPA		CAS			
Platanthera sparsiflora	RSA		CAS	RSA		
Platanthera stricta	RSA					
Spiranthes porrifolia	RSA					CAS
Spiranthes romanzoffiana	RSA		RSA			

Species	Lassen	Los Angeles	Madera	Marin	Mariposa	Mendocino
Calypso bulbosa				CAS		CAS
Cephalanthera austiniae			CAS		RSA	RSA
Corallorhiza maculata	RSA	RSA	RSA	RSA	RSA	RSA
Corallorhiza mertensiana						CAS
Corallorhiza striata	RSA		CAS	RSA	CAS	RSA
Corallorhiza trifida						
Cypripedium californicum				CAS		CAS
Cypripedium fasciculatum						
Cypripedium montanum			RSA		JEP/UC	CAS
Epipactis gigantea		CAS	CAS	RSA	RSA	RSA
Epipactis helleborine				CAS	RAC	
Goodyera oblongifolia			RSA	SFSU	RSA	RSA
Listera caurina						CAS
Listera convallarioides	RSA		OBI		CAS	
Listera cordata						PUA
Malaxis monophyllos						
Piperia candida						CAS
Piperia colemanii			UC		RSA	
Piperia cooperi		CAS				
Piperia elegans				CAS		JEP/UC
Piperia elongata		UCSB		CAS	RSA	CAS
Piperia leptopetala		UCSB				
Piperia michaelii		RSA		CAS		
Piperia transversa			JEP/UC	CAS	RSA	CAS
Piperia unalascensis	JEP/UC		CAS	CAS	RSA	CAS
Piperia yadonii						
Platanthera dilatata	CAS	RSA	RSA	RSA	RSA	RSA
Platanthera hyperborea					RSA	
Platanthera sparsiflora	RSA	RSA	CAS		RSA	CAS
Platanthera stricta						
Spiranthes porrifolia	CAS		RSA	CAS	CAS	RSA
Spiranthes romanzoffiana	CAS		CAS	CAS	CAS	CAS

Species	Merced	Modoc	Mono	Monterey	Napa	Nevada
Calypso bulbosa						
Cephalanthera austiniae				CAS		CAS
Corallorhiza maculata		US	RSA	CAS	RAC	CAS
Corallorhiza mertensiana						
Corallorhiza striata						CAS
Corallorhiza trifida						
Cypripedium californicum						
Cypripedium fasciculatum						CAS
Cypripedium montanum		RSA				
Epipactis gigantea		RSA	RSA	RSA	RSA	CAS
Epipactis helleborine				CAS	PUA	
Goodyera oblongifolia		JEP/UC				CAS
Listera caurina						
Listera convallarioides		CAS	JEP/UC			CAS
Listera cordata						
Malaxis monophyllos						
Piperia candida						
Piperia colemanii						
Piperia cooperi						
Piperia elegans				RSA	RAC	
Piperia elongata				CAS		RAC
Piperia leptopetala				CAS		JEP/UC
Piperia michaelii				CAS		
Piperia transversa				JEP/UC	CAS	CAS
Piperia unalascensis		CAS	UCSB	PGMNH		CAS
Piperia yadonii				JEP/UC		
Platanthera dilatata	HSC	CAS	RSA	RSA		RSA
Platanthera hyperborea			RSA			
Platanthera sparsiflora		CAS	RSA			RSA
Platanthera stricta		CAS				
Spiranthes porrifolia			UCSB	RSA	CAS	CAS
Spiranthes romanzoffiana		RSA	RSA	CAS		RSA

Species	Orange	Placer	Plumas	Riverside	Sacramento
Calypso bulbosa					
Cephalanthera austiniae		CAS	CAS		
Corallorhiza maculata		CAS	RSA	RSA	
Corallorhiza mertensiana			RSA		
Corallorhiza striata		SACT	RSA		
Corallorhiza trifida			CAS		
Cypripedium californicum			RSA		
Cypripedium fasciculatum			JEP/UC		
Cypripedium montanum			RSA		
Epipactis gigantea	RSA		RSA	RSA	
Epipactis helleborine					
Goodyera oblongifolia		CAS	RSA		
Listera caurina					
Listera convallarioides		JEP/UC	RSA	RSA	
Listera cordata					
Malaxis monophyllos				RSA	
Piperia candida					
Piperia colemanii		CAS	RSA		
Piperia cooperi	CAS			CAS	
Piperia elegans					
Piperia elongata			RSA	UCR	
Piperia leptopetala	JEP/UC		CAS	CAS	
Piperia michaelii					
Piperia transversa		CAS	RSA	RSA	
Piperia unalascensis		CAS	CAS		
Piperia yadonii					
Platanthera dilatata		RSA	RSA	RSA	
Platanthera hyperborea					
Platanthera sparsiflora		CAS	RSA	RSA	
Platanthera stricta					
Spiranthes porrifolia		RSA	CAS		
Spiranthes romanzoffiana		JEP/UC	RSA	RSA	

Species	San Benito	San Bernardino	San Diego	San Francisco	San Joaquin	San Luis Obispo
Calypso bulbosa						
Cephalanthera austiniae		UCR	SD			
Corallorhiza maculata		RSA	RSA			
Corallorhiza mertensiana						
Corallorhiza striata						
Corallorhiza trifida						
Cypripedium californicum						
Cypripedium fasciculatum						
Cypripedium montanum						
Epipactis gigantea	CAS	RSA	RSA			RSA
Epipactis helleborine				CAS		
Goodyera oblongifolia						
Listera caurina						
Listera convallarioides		RSA				
Listera cordata						
Malaxis monophyllos		RSA				
Piperia candida						
Piperia colemanii						
Piperia cooperi		CAS	RSA			
Piperia elegans	RAC			RSA		UCR
Piperia elongata	RAC	CAS	JEP	RSA		CAS
Piperia leptopetala	CAS	UCR	CAS			CAS
Piperia michaelii				CAS		RSA
Piperia transversa		RSA	CAS			CAS
Piperia unalascensis		RSA	JEP			
Piperia yadonii						
Platanthera dilatata		RSA	RSA			RSA
Platanthera hyperborea						
Platanthera sparsiflora		RSA				
Platanthera stricta						
Spiranthes porrifolia		RSA	CAS	JEP/UC	RSA	
Spiranthes romanzoffiana	CAS	RSA		RSA		RSA

Species	San Mateo	Santa Barbara	Santa Clara	Santa Cruz	Shasta	Sierra
Calypso bulbosa	RAC			RSA	CSU	
Cephalanthera austiniae				CAS	CAS	JEP/UC
Corallorhiza maculata	RSA		CAS	RSA	RSA	RSA
Corallorhiza mertensiana						
Corallorhiza striata	RSA		CAS	CAS		RAC
Corallorhiza trifida						
Cypripedium californicum					CAS	
Cypripedium fasciculatum	CAS		RSA	CAS		JEP/UC
Cypripedium montanum	CAS			CAS	CAS	
Epipactis gigantea	CAS	RSA	RSA	CAS	CAS	RSA
Epipactis helleborine	CAS		CAS	RAC		
Goodyera oblongifolia	RAC		CAS		JEP/UC	RAC
Listera caurina						
Listera convallarioides					CAS	CAS
Listera cordata						
Malaxis monophyllos						
Piperia candida	CAS			JEP/UC		
Piperia colemanii					CAS	
Piperia cooperi						
Piperia elegans	CAS	SBBG	CAS	JEP/UC		
Piperia elongata	RSA	CAS	SBBG	CAS	RAC	
Piperia leptopetala			CAS		NYBG	
Piperia michaelii	CAS	CAS	CAS	JEP/UC		
Piperia transversa	CAS		CAS	CAS	LA	RAC
Piperia unalascensis	US		CAS	CAS	CAS	CHSC
Piperia yadonii						
Platanthera dilatata	RSA		US		RSA	RSA
Platanthera hyperborea						
Platanthera sparsiflora					CAS	CAS
Platanthera stricta						
Spiranthes porrifolia					RSA	RAC
Spiranthes romanzoffiana	RSA			RSA	CAS	CAS

Species	Siskiyou	Solano	Sonoma	Stanislaus	Sutter	Tehama
Calypso bulbosa	RSA		RSA			
Cephalanthera austiniae	RSA		ROPA			RSA
Corallorhiza maculata	RSA		CAS			CAS
Corallorhiza mertensiana	RSA		CAS			
Corallorhiza striata	RSA		CAS			CAS
Corallorhiza trifida						
Cypripedium californicum	RSA		CAS			
Cypripedium fasciculatum	CAS					RSA
Cypripedium montanum	RSA		RSA			JEP/UC
Epipactis gigantea	CAS		RSA			CSU
Epipactis helleborine						
Goodyera oblongifolia	RSA		CAS			RSA
Listera caurina	JEP/UC					
Listera convallarioides	RSA					RSA
Listera cordata						
Malaxis monophyllos						
Piperia candida	JEP/UC		JEP/UC			
Piperia colemanii	JEP/UC					
Piperia cooperi						
Piperia elegans			CAS			
Piperia elongata	CAS		CAS			RSA
Piperia leptopetala	JEP/UC		RAC			
Piperia michaelii				JEP/UC		
Piperia transversa	CAS		CAS			CAS
Piperia unalascensis	RSA		UCR			CAS
Piperia yadonii						
Platanthera dilatata	RSA		RSA			RSA
Platanthera hyperborea	JEP/UC					
Platanthera sparsiflora	RSA					CAS
Platanthera stricta	CAS					
Spiranthes porrifolia	RSA		CAS	JEP/UC		CSU
Spiranthes romanzoffiana	RSA		CAS			RSA

Species	Trinity	Tulare	Tuolumne	Ventura	Yolo	Yuba
Calypso bulbosa	CAS					
Cephalanthera austiniae	RSA	RSA	CAS			SBBG
Corallorhiza maculata	RSA	RSA	CAS			RAC
Corallorhiza mertensiana	RSA					
Corallorhiza striata	CAS		CAS			
Corallorhiza trifida						
Cypripedium californicum	CAS					
Cypripedium fasciculatum	CAS					RAC
Cypripedium montanum	RSA		CAS			
Epipactis gigantea	CAS	RSA	RSA	RSA	RSA	SFSU
Epipactis helleborine						
Goodyera oblongifolia	RSA	RAC	CAS			RAC
Listera caurina						
Listera convallarioides	RSA	CAS	CAS			RAC
Listera cordata						
Malaxis monophyllos						
Piperia candida	CAS					
Piperia colemanii		CAS	CAS			
Piperia cooperi				LA		
Piperia elegans						
Piperia elongata	CAS	US	UOP	CAS		
Piperia leptopetala		CAS				
Piperia michaelii		RSA	CAS	CAS		CAS
Piperia transversa	CAS	CAS	CAS			
Piperia unalascensis	RSA	RSA	CAS			
Piperia yadonii						
Platanthera dilatata	RSA	RSA	RSA			SFSU
Platanthera hyperborea	HSC	JEP/UC				
Platanthera sparsiflora	RSA	RSA	RSA	RSA		
Platanthera stricta	CAS					
Spiranthes porrifolia	RSA	RSA	RSA			PUA
Spiranthes romanzoffiana	RSA	RSA	RSA			

Appendix 2. Potential Range Extensions and New Taxa

The distribution maps in the text consider a series confirmed in a county only if (1) there is an herbarium record to document its presence or (2) the author has observed it growing in the county. Potentially, many of the species have slightly larger ranges than shown, simply because the herbarium records cannot reflect the totality of reality. In several cases there are local floras or U.S. Forest Service reports of species in additional counties unsupported by herbarium data. For some species the habitat in counties adjacent to where they are confirmed seems ideal for them, and there is thus a good probability they grow there. Part 1 of this Appendix summarizes those cases. Some but not all of the potential locations are mentioned in the species text. Part 2 lists species that are not confirmed in California, but may be here because they occur in adjacent states in habitats similar to habitat that exists here. The intent of this Appendix is to challenge the reader to add to the known distribution of our wild orchids. Numbered footnotes give the reasons for the speculations where they are not simple habitat expansions.

Part 1. Potential new counties for species documented in California

Species	Potential or unconfirmed county locations
Calypso bulbosa	Napa[1]
Cephalanthera austiniae	Lassen, Modoc, Napa[1]
Corallorhiza maculata	Alpine
C. mertensiana	Nevada, Placer, Sierra
C. striata	Alpine, Del Norte, Modoc, San Diego,[2] Shasta
Cypripedium californicum	Butte,[3] Marin,[4] Sierra
C. montanum	Sierra, Nevada, Placer, El Dorado, Amador, Calaveras
Epipactis gigantea	Lassen, Sacramento, Sutter, San Joaquin, Stanislaus, Merced, King, Solano
E. helleborine	many[5]
Goodyera oblongifolia	Lassen
Listera caurina	Trinity
L. cordata	Trinity, Siskiyou
Malaxis monophyllos var. *brachypoda*	Riverside[6]
Piperia candida	Marin
P. colemanii	Sierra, Nevada, El Dorado, Amador, Calaveras
P. elongata	Orange, Sierra, Placer, Madera, Fresno
P. michaelii	Mendocino, Sonoma
P. unalascensis	Riverside
Platanthera hyperborea	Modoc, Shasta
P. stricta	Shasta
Spiranthes porrifolia	San Mateo, Santa Cruz
Part 2. Potential new species for California	
Cypripedium parviflorum var. *makasin*	Sierra,[7] Siskiyou,[8] Shasta,[8] Modoc[8]
Spiranthes infernalis	Inyo,[9] San Bernardino[9]

[1]Listed in local flora.

[2]Single herbarium record, perhaps erroneous.

[3]U.S. Forest Service field report.

[4]Documented in Marin County by herbarium records, but the only location noted was destroyed by storm.

[5]Non-native orchid with demonstrated ability to spread readily.

[6]Well-documented historical location, but multiple recent searches have failed to verify that the plant is still extant there.

[7]A single herbarium specimen of this species was collected, its annotation states, in Sierra County, but recent searches there did not verify occurrence.

[8]This species occurs in Oregon in areas where habitat is similar to that in these three counties.

[9]This species occurs in Nevada in habitat similar to parts of these two counties.

Glossary

acuminate tapering to a long, slender point at its apex
acute distinctly pointed, more abruptly than acuminate would imply
albino lacking all pigmentation
alternate having the leaves offset from one another (not across from one another) and on opposite sides of the stem
anterior on the front of, as of a flower or organ
anther the part of a stamen producing the pollen; in orchids, part of the column
anther cap the caplike structure covering the pollinia, usually (as in *Epipactis*) detachable
anther sac saclike enclosure for pollinia, found in *Platanthera* and *Piperia*
anthesis the period when a flower is fully opened
anthocyanin the substance producing the blue or red coloring in plants
apex the tip, as of an organ
apical at the apex or tip
ascending growing obliquely upward, or curved upward
asexual without sex, as where reproduction proceeds by the production of new leads from the parent plant, as in *Goodyera*
asymbiotic lacking a symbiont; with respect to orchids, used to refer to the germination of seeds in the laboratory without the presence of a compatible fungus
attenuate gradually tapering
autogamy self-pollination, without the aid of a vector
axis the central or main stem of a plant, in orchids usually elongated
basal at the base, as of an organ or part; said, for example, of leaves attached to the stem at ground level
bifurcate forked; two-parted
bilateral symmetry the form of symmetry expressed as two identical halves forming mirror images of each other; in orchids a line from the tip of the dorsal sepal to the tip of the lip bisects the flower into two identical halves
bilobed having two lobes
bract a reduced, leaflike structure attached below a flower or along the stem
callose having callosities; hard or thick

callosity a thickening, as of part of an organ
callus (pl. calli) a hard or solid protuberance
capsule the mature ovary of some flowers (for example of orchids) that splits when ripe and contains the seeds
caudicle an extension or tail to the pollinium that sticks to the pollinator or the viscidium
cauline attached to the stem at some point above the ground; said of leaves
chaparral a plant community typical of the dry hillsides of Southern California, consisting of shrubs and small trees and usually quite dense
character a feature of plants (in some group) used (usually in conjunction with others) to distinguish the several taxa constituting that group
chlorophyll the green pigment characteristic of plants; an essential compound in photosynthesis
clasping partly or wholly surrounding the stem or other organ, as a leaf or bract
clavate club-shaped, with the blade thicker toward the apex than at the base
coastal scrub a coastal plant community consisting of shrubs less than 2 meters tall, such as (in one such community) *Baccharis pilularis, Eriophyllum staechadifolium,* and *Rubus vitifolius*
colony a group of plants of the same species growing together
column the unique structure of orchids formed by the union of the pistil and stamens
concave curved inward
conduplicate folded together lengthwise
connate united
connivent converging or coming together but not united
convex curved outward
coralloid coral-like; said, for example, of the root mass of certain orchids
cordate heart-shaped; said usually of leaves having the base lobed on each side as in the top of the heart
corm the solid, thickened, upright base of a stem
cross-pollination an act of pollination in which pollen from one flower is deposited on the stigma of another flower, usually on a different plant
cuneate triangular or wedge-shaped, as the base of an organ that is attached at the narrow end
decumbent lying on the ground with the growing end ascending
decurrent extending down the stem or axis, as a leaf or inflorescence
dehisce to split into different parts along a line or slit, so as to discharge contents, as seeds from a capsule
deltoid broadly triangular
depauperate having less than normal vigor, or stunted; having relatively few of something, as leaves
dilate/dilated expanded or widened

disjunct separated, as a population of some plant occurring at considerable remove from the remaining distribution of that taxon
dissected cut into numerous segments
dorsal on the upper side, as of a flower; in orchids, usually used in reference to the uppermost sepal
duff forest-floor covering composed of decaying leaves, conifer needles, small branches, etc.
ellipsoid elliptic in longitudinal cross-section; said of a three-dimensional object, as many capsules
elliptic oval; widest in the middle and with rounded ends; said of an essentially two-dimensional object, as a leaf
endangered the California Endangered Species Act of 1984 declares a native plant "endangered when its prospects of survival and reproduction are in immediate jeopardy from one or more causes" (see also "rare" and "threatened")
endemic restricted to a given geographic region; occurring nowhere else in the world
entire without divisions, lobes, or teeth on its periphery, as a leaf or petal
epichile the terminal part of the labellum when it is distinct from the basal part, or hypochile
epithet the adjective, following the genus name, that forms the (second half of the) name of a species
falcate sickle-shaped
fen flat wetland having a groundwater source
filiform threadlike; long and slender
floral envelope the sepals and petals considered as a unit
fruit a ripened ovary
fusiform wide in the middle and narrow at both ends (spindle-shaped)
galea (pl. galeae) a hood or helmet formed (in orchids) by the sepals and petals
genus the taxonomic ranking above species, but below family; the genus name is used with the specific epithet to form the species name, or binomial
glabrous without hairs or scales
habitat the normal environmental situation in which a plant grows
hastate shaped like an arrowhead with the basal lobes turned outward, as a leaf
hemipollinarium (pl. hemipollinaria) one of a pair of viscidia with stipe or caudicle and pollinium from the same flower
herbaceous herblike; having no woody stems above ground
humus decomposing organic material in the soil
hybrid offspring of plants with different genetic material, especially of different species
hypochile the basal part of a complex labellum when it is distinct from the terminal part, or epichile
indigenous native to a region (but not to that region alone)

inferior ovary — an ovary bearing the flower parts at its apex
inflorescence — a cluster of one to many flowers on a single stem
labellum — the lip
lamina — the usually flat, expanded part of a leaf or floral segment
lanceolate — lance-shaped; long and tapering with a broad base
lateral — belonging to or borne on the side, as for example lateral lobes
linear — long and narrow, usually with little or no taper along the length
lingulate — tongue-shaped
lip — the labellum; a modified petal usually differing from the other petals in size, shape, and sometimes color
lobate — having lobes
lobe — a segment of a dissected organ, as a leaf blade, especially if rounded
mentum — a chinlike or spurlike projection formed by the extension (in orchids) of the bases of the lateral sepals
monocotyledon — one of two major divisions of angiosperms (the division that includes the orchids and others) characterized by a single embryonic leaf (rather than two) at germination
monotypic — having a single subordinated taxon, as a genus of one species
morphology — the form and structure of an organism; also the study of form, and of its evolution and development
mycorrhizal — having a symbiotic relationship between the roots (as of vascular plants) and a fungus
mycotrophic — obtaining food through a mycorrhizal relationship with a fungus; said of certain vascular plants
node — a place (or the place) on the stem that bears leaves, bracts, or other organs
oblanceolate — lanceolate, with the broadest part toward the apex
oblique — having unequal sides; slanting
oblong — several times longer than wide
obovate — two-dimensionally egg-shaped and attached at the narrow end
obovoid — three-dimensionally egg-shaped and attached at the narrow end
obtuse — pointed, with the sides meeting at an angle of greater than 90 degrees
ovary — the part of the pistil containing the ovules
ovate — two-dimensionally egg-shaped and attached at the broad end
pandurate — fiddle-shaped, often used in reference to the lip or petals of orchids
pedicel — the stalk between the flower and the floral bract
peduncle — the stalk of an inflorescence, or the stem of a solitary flower
peloria — an abnormal radial symmetry by mutation in a normally bilaterally symmetric flower; in peloric orchids, all three petals are shaped like lips
perennating — lasting the whole year through
perennial — a plant that lives more than two years
perfect — having functional pistil and stamens; in orchids, said of the anther
petal — one of three parts forming the inner whorl of the orchid flower; in orchids, one petal is modified to form a lip

phenology the study of the periodic phenomena of plants, such as leafing, flowering, fruiting, etc.

photosynthesis the process by which plants use incident light energy to produce carbohydrates from carbon dioxide and water

pistil the female, or seed-producing, organ of a flower; in orchids the pistil is part of the column

plicate folded lengthwise into plaits, as a fan

pollen the grains or mass of grains borne by the anther that develop the male gametes when they germinate in the stigma

pollinarium (pl. pollinaria) a unit consisting of pollinium and associated structures if present, e.g. stipe, viscidium, caudicle

pollinium (pl. pollinia) a coherent mass of pollen, as in orchids

polymorphic having more than one form, as leaves of different forms on the same plant

protandrous maturing before the pistils in the same flower; said of the anthers

raceme an unbranched inflorescence bearing pedicellate flowers

rare the California Endangered Species Act of 1984 declares a native plant "rare when, although not presently threatened with extinction, it is in such small numbers throughout its range that it may become endangered if its present environment worsens" (see also "endangered" and "threatened")

reflexed bent backward abruptly

resupinate rotated so as to be presented upside down; in orchids, referring to the twisting of the pedicel, usually while the flower is in the bud stage, that positions the lip at the lower side

reticulated having a network of veins, as a leaf, the veins usually of a different color than the leaf

rhizome a horizontal stem growing underground or along the substrate

rostellum a slender extension from the upper edge of the stigma in orchids

saccate shaped like a pouch or bag

saprophyte a plant that does not manufacture its own food, but depends on organic matter in the soil

scape a peduncle, supporting an inflorescence, that rises from the ground with or without bracts, but never with well-developed leaves

sepal one of the several floral parts seen most prominently in the bud stage, often similar to the petals, which are protected in the bud by the sepals; in orchids, one element of the outer whorl of the flower

sessile attached directly at the base, without a stalk

sheathing the base of a leaf or bract, wrapped around the stem

spike an unbranched, indeterminate inflorescence with sessile flowers

spur a hollow, tubular, or sacklike projection from a flower part, usually containing nectar; commonly (in orchids) a rearward extension of the lip

stamen male, or pollen-producing, organ of a flower; in orchids the stamen is part of the column staminode a sterile stamen, bearing no pollen; in *Cypripedium* the staminode is the flat structure that partially covers the opening to the pouch

stem the ascending axis (or one of the ascending axes) of the plant, usually bearing the leaves, flowers, and fruit

stigma (pl. stigmata) the part of the pistil, usually sticky, that receives the pollen

stipe a stalk; in some orchids the stalk that connects the viscidium with a pollinium

striated marked with lines or stripes of a different color

symbiont an organism that lives cooperatively with, and benefits from, an organism of a different species that itself benefits from the association

sympatric occupying the same or an overlapping geographic range without interbreeding

synsepal a sepal formed from the uniting of two lateral sepals; found in *Cypripedium*

taxon a unit or category in taxonomic classification, as a family or species

taxonomy the science of classification of plants or animals

terrestrial growing on the ground, as contrasted with epiphytic, or growing on trees

threatened the California Endangered Species Act of 1984 declares a native plant threatened when although not presently facing extinction, it is "likely to become an endangered species in the foreseeable future in the absence of special protection and management efforts" (see also "rare" and "endangered")

translucent transmitting light, but with some diffusion, as some petals or sepals

tuber a thickened or shortened underground stem capable of vegetative reproduction

tuberosity a projection or protuberance from an organ; used here as a small projection from the lip

undulate wavy

unifoliate having a single leaf

variety an infraspecific taxon (one of two or more making up a species) differing in more or less minor ways from the typical

viscidium (pl. viscidia) a sticky structure usually connected, in some orchids, to the caudicles or stipes that carry the pollen masses, thus aiding in the attachment of the pollen to pollinators

whorl three or more leaves, sepals, or petals arranged in a circle about an axis

Bibliography

This Bibliography lists all the references cited in the text except those that are the sources of species or genus descriptions (cited under the primary text headings) or synonyms (cited under "synonymy"). In addition, several excellent reference sources cited nowhere in text (works on orchids in general, local and regional floras, orchid books for states other than California, and articles on specific species or genera) are also listed, as an aid to the reader interested in additional data or pursuing further research. Works not cited in the text are preceded by asterisks.

Abrams, L. 1904. *Flora of Los Angeles and Vicinity.* Stanford University Press, Stanford, Calif., 93–95.

Abrams, L. 1940. *Illustrated Flora of the Pacific States,* Vol. 1. Stanford University Press, Stanford, Calif., 469–484.

Ackerman, J. D. 1975. Reproductive Biology of *Goodyera oblongifolia* (Orchidaceae). *Madroño* 23 (4): 191–198.

Ackerman, J. D. 1977. Biosystematics of the Genus *Piperia* Rydb. *Botanical Journal of the Linnean Society* 75: 245–270.

Ackerman, J. D. 1981. Pollination Biology of *Calypso bulbosa* var. *occidentalis* (Orchidaceae): A Food Deceptive System. *Madroño* 28 (3): 101–110.

Ackerman, J. D., and M. R. Mesler. 1979. Pollination Biology of *Listera cordata* (Orchidaceae). *American Journal of Botany* 66 (7): 820–824.

*Allen, D. 1982. *Epipactis gigantea. American Orchid Society Bulletin* 51 (10): 1038–1040.

Ames, O. 1910. *Orchidaceae,* Vol. IV: *The Genus Habenaria in North America.* The Merrymount Press, Boston [reprinted in 1979].

Ames, O., and D. Correll. 1943. Notes on North American Orchids. *Botanical Museum Leaflets, Harvard University* 11 (1): 1–2.

*Arditti, J. 1992. *Fundamentals of Orchid Biology.* John Wiley & Sons, New York.

Arditti, J., A. Oliva, and J. Michaud. 1982. Practical Germination of North American and Related Orchids, 2: *Goodyera oblongifolia* and *G. tesselata. American Orchid Society Bulletin* 51 (4): 394–397.

*Arditti, J., A. Oliva, and J. Michaud. 1985. Practical Germination of North American and Related Orchids, 3: *Calopogon tuberosus, Calypso bulbosa, Cypripedium* Species and Hybrids, *Piperia elegans* var. *elata, Piperia maritima, Platanthera hyperborea,* and *Platanthera saccata. American Orchid Society Bulletin* 54 (7): 859–866.

Atwood, J. T. 1984. The Relationships of the Slipper Orchids (Subfamily Cypripedioideae). *Selbyana* 7 (2, 3, 4): 129–247.

Barbour, M. G., and J. Major, eds. 1988. *Terrestrial Vegetation of California.* University of California Press, Berkeley.

Bartlett, H. H. 1922. Color Types of *Corallorhiza maculata* Raf. *Rhodora* 24: 145–148.

Bartlett, H. H. 1925. The Varieties of *Corallorhiza maculata. Rhodora* 27: 11–14.

*Beauchamp, R. M. 1986. *A Flora of San Diego County, California.* Sweetwater Press, National City, Calif.

*Biek, D. 1988. *Flora of the Whiskeytown National Recreation Area, Shasta County, California.* U.S. National Park Service.

Bingham, M. T. 1939. *Orchids of Michigan.* Cranbrook Institute of Science, Bloomfield Hills, Mich.

Brownell, V. R., and P. M. Catling. 1987. Notes on the Distribution and Taxonomy of *Cypripedium fasciculatum* Kellogg ex Watson. *Lindleyana* 2 (1): 53–57.

Brunton, D. F. 1986. Status of the Giant Helleborine, *Epipactis gigantea* (Orchidaceae), in Canada. *Canadian Field-Naturalist* 100 (3): 414–417.

Calder, J. A., and R. L. Taylor. 1968. *Flora of the Queen Charlotte Islands.* Research Branch, Canadian Dept. of Agriculture, 287–300.

*Cameron, J. W. 1976. *The Orchids of Maine.* University of Maine at Orono Press.

*Carville, J. S. 1989. *Lingering in Tahoe's Wild Gardens.* Mountain Gypsy Press, Chicago Park, Calif.

Case, F. W. 1987. *Orchids of the Western Great Lakes Region.* Cranbrook Institute of Science Bulletin.

Case, F. W. 1988. The Jewel Orchids of North America. *American Orchid Society Bulletin* 57 (7): 758–765.

Catling, P. M., and V. Catling. 1991. Synopsis of Breeding Systems and Pollination in North American Orchids. *Lindleyana* 6 (4): 187–210.

Catling, P. M., and C. J. Sheviak. 1993. Taxonomic Notes on Some North American Orchids. *Lindleyana* 8 (2): 77–81.

*Clemons, D. 1986. *Plants of Anza-Borrego Desert State Park.* Anza-Borrego Desert Natural History Association, Borrego Springs, Calif.

*Cockerell, T. D. 1903. Two Orchids from New Mexico. *Torreya* 3: 139–141.

*Cockerell, T. D. 1916. A New Form of *Corallorhiza. Torreya* 16: 230–232.

*Coleman, R. A. 1986. The Habitat and Variation of *Epipactis gigantea* in the Santa Monica Mountains. *Orchid Digest* 50 (2): 86–87.

*Coleman, R. A. 1988a. The Coral Root Orchids of California. *Fremontia* 16 (3): 21–22.

*Coleman, R. A. 1988b. The *Epipactis* of California. *Fremontia* 16 (1): 24–27.

*Coleman, R. A. 1988c. The Orchids of Yosemite National Park. *American Orchid Society Bulletin* 57 (6): 609–623.

*Coleman, R. A. 1989a. Stalking *Calypso bulbosa* in the American West. *American Orchid Society Bulletin* 58 (10): 1023–1028.

*Coleman, R. A. 1989b. *Cypripediums* of California. *American Orchid Society Bulletin* 58 (5): 456–460.

*Coleman, R. A. 1989c. *Epipactis helleborine* on the West Coast. *Orchid Digest* 53 (2): 84–86.

Coleman, R. A. 1989d. *Listeras*, Some Overlooked Orchids. *Fremontia* 17 (3): 26–27.

*Coleman, R. A. 1989e. Orchids in California's Coastal Scrub. *American Orchid Society Bulletin* 58 (7): 661–665.

*Coleman, R. A. 1990a. *Malaxis monophyllos* var. *brachypoda* Rediscovered in California. *American Orchid Society Bulletin* 59 (1): 41–45.

*Coleman, R. A. 1991a. A Disjunct Location for *Corallorhiza trifida* var. *verna. American Orchid Society Bulletin* 60 (4): 329–330.

*Coleman, R. A. 1991b. A Fourth Coral Root in California. *Fremontia* 19 (1): 22–23.

*Coleman, R. A. 1991c. Blooming Seasons for Wild Orchids in California. *American Orchid Society Bulletin* 60 (9): 876–879.

*Coleman, R. A. 1991d. The Orchid Genus *Platanthera* in California. *Fremontia* 19 (2): 19–22.

*Coleman, R. A. 1992. Friendly Fire. *American Orchid Society Bulletin* 61 (2): 130–135.

*Correll, D. S. 1943. The Genus *Habenaria* in Western North America. *Leaflets of Western Botany* III: 233–256.

Correll, D. S. 1978. *Native Orchids of North America.* Stanford University Press, Stanford, Calif. [reprint of 1950 book].

Crandall, C. A. 1900. The Tall Green Orchis (*Habenaria hyperborea*) Visited by Mosquitoes. *Plant World* 1: 6–7.

*Cronquist, A., A. Holmgren, N. Holmgren, J. Reveal, and P. Holmgren. 1977. *Intermountain Flora,* Six: *The Monocotyledons.* Columbia University Press, New York, 546–566.

Darwin, C. 1877. *The Various Contrivances by Which Orchids are Fertilised by Insects*. University of Chicago Press, Chicago [1984 reprint].

Davies, P., J. Davies, and A. Huxley. 1988. *Wild Orchids of Britain and Europe*. Hogarth Press, London.

Dressler, R. L. 1981. *The Orchids, Natural History and Classification*. Harvard University Press, Cambridge, Mass.

Dressler, R. L. 1993. *Phylogeny and Classification of the Orchid Family*. Deoscorides Press, Portland.

Dunsterville, G. C. K., and E. Dunsterville. 1986. Orchids—What They Are, and What They Do. *American Orchid Society Bulletin* 55 (6): 604–611.

Eastman, E. C. 1990. *Rare and Endangered Plants of Oregon*. Beautiful America Publishing Company, Wilsonville, Oregon, 53–58.

Farwell, O. A. 1923. *Corallorhiza maculata*. *Rhodora* 25: 31–32.

*Ferlatte, W. J. 1974. *A Flora of the Trinity Alps of Northern California*. University of California Press, Berkeley.

*Fisher, R. M. 1980. *Guide to the Orchids of the Cypress Hills*. Fisher.

*Fowlie, J. A. 1980. The Occurrence of Ladyslipper Orchids and Jewel Orchids, Part II: The Occurrence of *Cypripedium montanum* in Yosemite National Park with *Goodyera oblongifolia*. *Orchid Digest* 44 (1): 19–22.

*Fowlie, J. A. 1982. Notes on the Habitat and Ecological Relationship of *Cypripedium californicum* A. Gray and *Darlingtonia californica*. *Orchid Digest* 46 (5): 164–170.

Freudenstein, J. V., and L. H. Bailey. 1987. A Preliminary Study of *Corallorhiza maculata* (Orchidaceae) in Eastern North America. *Contributions University of Michigan Herbarium* 16: 145–153.

Fries, M. A. 1970. *Wildflowers of Mount Rainier and the Cascades*. The Mountaineers, Seattle.

Fultz, F. M. 1928. *Lily, Iris, and Orchid of Southern California*. Spanish American Institute Press, Gardena, Calif., 127–133.

Garay, L. S. 1980. A Generic Revision of the Spiranthinae. *Botanical Museum Leaflets, Harvard University* 28 (4): 278–425.

*Gillett, G. W., J. Howell, and H. Leschke. 1961. A Flora of Lassen Volcanic National Park, California. *Wasmann Journal of Biology* 19 (1): 1–66.

Godfery, M. J. 1922. *Spiranthes romanzoffiana*. *Orchid Review*: 261–264.

*Gray, A. 1876. Contributions to the Botany of North America. *Proceedings of the American Academy of Arts and Sciences* 12: 83.

Gray, A. 1879. *Epipactis helleborine* var. *viridens*. *Botanical Gazette* IV (9): 206.

Greene, E. L. 1895. Novitates Occidentales, XI. *Erythea* 3: 49.

Griesbach, R. J. 1979. The Albino Form of *Epipactis helleborine*. *American Orchid Society Bulletin* 48 (8): 808–809.

*Gupton, O. W., and F. C. Swope. 1987. *Wild Orchids of the Middle Atlantic States*. University of Tennessee Press, Knoxville.
*Hall, H. M. 1902. *A Botanical Survey of San Jacinto Mountain*. University of California Publications, Vol. 1, 69–71.
*Harvais, G. 1974. Notes on the Biology of Some Native Orchids of Thunder Bay, Their Endophytes and Symbionts. *Canadian Journal of Botany* 52: 451–460.
Haskin, L. 1967. *Wildflowers of the Pacific Coast*. Bindfords and Mort, Portland.
*Hawkes, A. D. 1965. *Encyclopaedia of Cultivated Orchids*. Faber and Faber Ltd., London.
Heller, A. A. 1898. New Plants from Western North America. *Torrey Botanical Club* 25: 193.
Heller, A. A. 1900. *Catalogue of North American Plants North of Mexico:* 4.
Heller, A. A. 1904. Western Species, New and Old, II. *Muhlenbergia* 1 (4): 48–49.
*Henrich, J. E., D. Stimart, and P. Archer. Terrestrial Orchid Seed Germination in Vitro on a Defined Medium. *Journal American Horticultural Society* 106 (2): 193–196.
Hickman, J. C., ed. 1993. *The Jepson Manual: Higher Plants of California*. University of California Press, Berkeley.
*Higgins, W. E. 1989. The Wild Orchids of the Central Southwest. *American Orchid Society Bulletin* 58 (7): 666–670.
Hill, M. 1984. *California Landscapes*. University of California Press, Berkeley.
Hill, R. B. 1986. *California Mountain Ranges*. Falcon Press, Helena, Mont.
Hitchcock, C. L. 1969. *Vascular Plants of the Pacific Northwest*. University of Washington Press, Seattle.
*Holing, D. 1988. *California Wild Lands*. Chronicle Books, San Francisco.
*Hoover, R. F. 1970. *The Vascular Plants of San Luis Obispo County, California*. University of California Press, Berkeley.
Howell, J. T. 1970. *Marin Flora*. University of California Press, Berkeley.
Howell, J. T., P. H. Raven, and P. Rubtzoff. 1958. A Flora of San Francisco. *Wasmann Journal of Biology* 16 (1): 1.
*Howitt, B. F., and J. T. Howell. 1964. The Vascular Plants of Monterey County, California. *Wasmann Journal of Biology* 22 (1): 158.
*Jackson, B. D. 1971. *A Glossary of Botanic Terms,* fourth edition. Duckworth and Company, London [reprint of 1928 edition].
Jepson, W. L. 1951. *A Manual of the Flowering Plants of California*. University of California Press, Berkeley.
Jorgensen, E. 1982. *Epipactis helleborine* of Milwaukee's Lake Park—Revisited After Fifty Years. *American Orchid Society Bulletin* 51 (1): 41–42.
Judd, W. W. 1971. Wasps Pollinating Helleborine, *Epipactis helleborine* (L.) Crantz, at Owens Sound, Ontario. *Proceedings, Entomological Society of Ontario* 102: 115–118.

*Keenan, P. E. 1983. *A Complete Guide to Maine's Orchids.* Delorme Publishing Company, Freeport, Maine.

*Keenan, P. E. 1988. *Calypso bulbosa:* Hider of the North. *American Orchid Society Bulletin* 57 (4): 373–377.

Kipping, J. L. 1971. Pollination Studies of Native Orchids. M.S. thesis, San Francisco State College [unpublished].

*Knight, W., I. Knight, and J. T. Howell. 1970. A Vegetation Survey of the Butterfly Botanical Area, California. *Wasmann Journal of Biology* 28 (1): 1–19.

Küchler, A. W. 1988. The Map of the Natural Vegetation of California. In: M. Barbour and J. Major, eds., *Terrestrial Vegetation of California.* California Native Plant Society, Sacramento.

Larson, K. S., and R. J. Larson, 1990. Lure of the Locks: Showiest Ladies-tresses Orchids, *Spiranthes romanzoffiana,* Affect Bumblebee, *Bombus* ssp., Foraging Behavior. *Canadian Field-Naturalist* 104 (4): 519–525.

Larson, R. J. 1992. Pollination of *Platanthera dilatata* var. *dilatata* in Oregon by the Noctuid Moth *Discestra oregonica. Madroño* 39 (3): 236–237.

Larson, R. J., and K. S. Larson. 1987. Observations on the Pollination Biology of *Spiranthes romanzoffiana. Lindleyana* 2 (4): 176–179.

*Lathrop, E. W., and R. F. Thorne. 1978. A Flora of the Santa Ana Mountains, California. *Aliso* 9 (2): 197–278.

*Lathrop, E. W., and R. F. Thorne. 1985. A Flora of the Santa Rosa Plateau, Southern California. *Southern California Botanist,* Special Publication #1: 269.

Lichvar, R. W. 1979. *Epipactis gigantea* Douglas ex Hook. *Madroño* 26: 188.

Light, M., and M. MacConaill. 1989. Albinism in *Platanthera hyperborea. Lindleyana* 4 (3): 158–160.

Light, M., and M. MacConaill. 1990. Population Dynamics of a Terrestrial Orchid. In: *Proceedings of the 13th World Orchid Conference.* 1990 World Orchid Conference Trust, Auckland, N.Z., 245–247.

Light, M., and M. MacConaill. 1991. Patterns of Appearance in *Epipactis helleborine* (L.) Crantz. In T. C. E. Wells and J. H. Willems, eds. *Population Ecology of Terrestrial Orchids.* SPB Publishing bv, The Hague, 77–87.

*Lindley, J. 1830. *The Genera and Species of Orchidaceous Plants.* Ridgeway, London.

*Lloyd, R. M., and R. S. Mitchell. 1973. *A Flora of the White Mountains of California and Nevada.* University of California Press, Berkeley.

Long, J. C. 1970. *Native Orchids of Colorado.* Museum Pictorial 16, Denver Museum of Natural History.

Long, R. 1979. *Eburophyton austiniae* (A. Gray) A. A. Heller. *Davidsonia* 10 (2): 30–33.

Long, R. 1980. *Calypso bulbosa* (Linnaeus) Oakes in Z. Thompson. *Davidsonia* 11 (1): 13–16.

*Luer, C. A. 1969. The Genus *Cypripedium* in North America. *American Orchid Society Bulletin* 38: 903–908.

Luer, C. A. 1972. *The Native Orchids of Florida.* New York Botanical Garden, New York.

Luer, C. A. 1975. *The Native Orchids of the United States and Canada.* The New York Botanical Garden, New York.

MacDougal, D. T. 1899. Symbiosis and Saprophytism. *Bulletin of the Torrey Botanical Club* 26 (10): 511–529.

Major, J. 1988. California Climate in Relationship to Vegetation. In: M. Barbour and J. Major, eds., *Terrestrial Vegetation of California.* California Native Plant Society, Sacramento.

*McAulay, M. 1985. *Wildflowers of the Santa Monica Mountains.* Canyon Publishing Company, Canoga Park, Calif.

McClintock, E. 1975. The Travels of an Orchid. *California Horticultural Journal* 36 (2): 52–53.

Morgan, R., and J. Ackerman. 1990. Two New Piperias (Orchidaceae) from Western North America. *Lindleyana* 5 (4): 205–211.

Morgan, R., and L. Glicenstein. 1993. Additional California Taxa in *Piperia* (Orchidaceae). *Lindleyana* 8 (2): 89–95.

Morris, F., and A. E. Eames. 1929. *Our Wild Orchids.* Charles Scribner's Sons, New York.

*Mosquin, T. 1970. The Reproductive Biology of *Calypso bulbosa* (Orchidaceae). *Canadian Field-Naturalist* 84: 291–296.

Mousley, H. 1927. The Genus *Amesia* in North America. *Canadian Field-Naturalist* 41 (1): 28–31.

Mousley, H. 1944. Peloria and Other Abnormalities in Orchids. *Canadian Field-Naturalist* 78: 73–76.

Munz, P. A. 1935. *Manual of Southern California Botany.* University of California Press, Berkeley.

Munz, P. A. 1968. *A Flora of California and Supplement.* University of California Press, Berkeley, 1394–1400.

Myers, P., and P. Ascher. 1982. Culture of North American Orchids from Seed. *HortScience* 17 (4): 550.

*Niehaus, T. F. 1974. *Sierra Wildflowers.* University of California Press, Berkeley.

Olive, P., and J. Arditti. 1984. Seed Germination of North American Orchids, II: Native California and Related Species of *Aplectrum, Cypripedium,* and *Spiranthes. Botanical Gazette* 145 (4): 495–501.

*Parrish, S. B. 1917. Plants of the San Bernardino Mountains. *Plant World* 20: 209.

Parsons, M. E. 1907. *The Wild Flowers of California.* Cunningham, Curtiss, and Welch, San Francisco.

Patt, J. M., M. Merchant, D. Williams, and B. Meeuse. 1989. Pollination Biology of *Platanthera stricta* (Orchidaceae) in Olympic National Park. *American Journal of Botany* 76 (8): 1097–1106.

Petrie, W. 1981. *Guide to the Orchids of North America.* Hancock House, Blaine, Wash.

Ramsey, C. T. 1950. The Triggered Rostellum of the Genus *Listera. American Orchid Society Bulletin* 19: 482–485.

*Raven, P. H. 1988. The California Flora. In: M. Barbour and J. Major, eds., *Terrestrial Vegetation of California.* California Native Plant Society, Sacramento, 109–138.

Raves, P., H. Thompson, and B. Prigge. 1986. *Flora of the Santa Monica Mountains.* University of California Press, Berkeley.

Rondeau, J. H. 1991. *Carnivorous Plants of California.* Rondeau, San Jose, Calif.

Ross, E. S. 1988. Does *Epipactis gigantea* Mimic Aphids? *Fremontia* 16 (2): 28–29.

Rydberg, P. A. 1901. The American Species of *Limnorchis* and *Piperia,* North of Mexico. *Bulletin of the Torrey Botanical Club* 28: 605–643.

Salmia, A. 1986. Chlorophyll-free Form of *Epipactis helleborine* (Orchidaceae) in SE Finland. *Annales Botanici Fennici* 23: 49–57.

Salmia, A. 1989. General morphology and anatomy of chlorophyll-free and green forms of *Epipactis helleborine* (Orchidaceae). *Annales Botanici Fennici* 26: 95–105.

*Sanders, D. J. 1978. *Epipactis helleborine*—A European Import. *American Orchid Society Bulletin* 47 (5): 426–427.

Scheffer, V. 1970. Phantom Orchid. *Pacific Discovery* 23: 30–31.

*Schrenk, W. J. 1975. Zum Problem der Variabilitat Innerbalb der Gattung *Platanthera* L. C. Rich. (*Habenaria* Willd. s.l.p.p.) in Nordamerika. *Die Orchidee* 26: 258–263.

*Schrenk, W. J. 1978. North American *Platantheras:* Evolution in the Making. *American Orchid Society Bulletin* 47 (3): 429–437.

*Sharsmith, H. K. 1982. *Flora of the Mount Hamilton Range of California.* California Native Plant Society, Sacramento.

Sheehan, J. 1992. *Goodyera* Propagation. *American Orchid Society Bulletin* 61 (9): 892.

*Sheviak, C. J. 1974. *An Introduction to the Ecology of the Illinois Orchidaceae.* Illinois State Museum, Springfield.

Sheviak, C. J. 1982. Biosystematic Study of the *Spiranthes cernua* Complex. New York State Museum, Bulletin 448, Albany.

Sheviak, C. J. 1984. *Spiranthes diluvialis* (Orchidaceae): A New Species from the Western United States. *Brittonia* 36: 8–14.

Sheviak, C. J. 1989. A New *Spiranthes* from Ash Meadows, Nevada. *Rhodora* 91: 225–234.

*Sheviak, C. J. 1990a. A New Form of *Cypripedium montanum* Dougl. ex Lindl. *Rhodora* 92 (870): 47–49.

Sheviak, C. J. 1990b. A New *Spiranthes* (Orchidaceae) from the Cienegas of Southernmost Arizona. *Rhodora* 92 (872): 213–229.

Sheviak, C. J. 1993. *Cypripedium parviflorum* Salib. var. *makasin* (Farwell) Sheviak. *American Orchid Bulletin* 62 (4): 403.

*Smith, C. F. 1976. *A Flora of the Santa Barbara Region, California.* Santa Barbara Museum of Natural History, Santa Barbara, Calif., 107.

*Smith, G. L. 1962. *Flowers of Lassen.* Loomis Museum Association, Lassen National Park, Calif.

*Smith, G. L. 1984. *A Flora of the Tahoe Basin and Neighboring Areas and Supplement.* The University of San Francisco, San Francisco, Calif.

Smith, J. P., and K. Berg. 1988. *California Native Plant Society's Inventory of Rare and Endangered Vascular Plants of California.* California Native Plant Society, Sacramento.

Smith, J. P., and R. York. 1984. *Inventory of Rare and Endangered Vascular Plants of California.* California Native Plant Society, Sacramento.

Sorrie, B. A. 1978. *Corallorhiza trifida* in California and Nevada. *Wasmann Journal of Biology* 36 (1&2): 199–200.

*State of California, Department of Fish and Game. 1987. Designated Endangered, Threatened, or Rare Plants. [leaflet]

Stoutamire, W. P. 1964. Seeds and Seedlings of Native Orchids. *The Michigan Botanist* 3: 107–119.

*Stoutamire, W. P. 1967. Flower Biology of the Lady's-Slippers. *The Michigan Botanist* 6: 159–175.

*Stoutamire, W. P. 1981. Early Growth in North American Terrestrial Seedlings. In: E. Plaxton, ed., *North American Terrestrial Orchids.* Michigan Orchid Society, Southfield, 14–24.

Summerhayes, V. S. 1968. *Wild Orchids of Britain.* Collins, London.

Szczawinski, A. F. 1975. *The Orchids of British Columbia.* British Columbia Provincial Museum, Victoria, Canada.

*Thomas, J. H. 1961. *Flora of the Santa Cruz Mountains of California.* Stanford University Press, Stanford, Calif., 129–131.

Todsen, T. A., and T. K. Todsen. 1971. Color Variation of *Corallorhiza* in New Mexico. *Southwestern Naturalist* 16 (1): 121–122.

van der Pijl, L., and C. Dodson. 1966. *Orchid Flowers, Their Pollination and Evolution.* University of Miami Press, Coral Gables, Fla.

Watson, J., ed. 1992. Caught in the Act. *American Orchid Society Bulletin* 61 (12): 1250.

Watson, S. 1876. Botanical Contributions 3. *Proceedings of the American Academy of Arts and Sciences* 11: 147.

Whiting, R. E., and P. M. Catling. 1986. *Orchids of Ontario*. The Canacoll Foundation, Ottawa, Canada.

Wiegand, K. M. 1899. A Revision of the Genus *Listera*. *Bulletin of the Torrey Botanical Club* 26 (4): 157–171.

Wilken, D. H., and W. F. Jennings. 1993. Orchidaceae. In: J. C. Hickman, ed., *The Jepson Manual: Higher Plants of California*. University of California Press, Berkeley.

*Williams, J. G., and A. E. Williams. 1983. *Field Guide to the Orchids of North America*. Universe Books, New York.

*Williams, L. O. 1937. The Orchidaceae of the Rocky Mountains. *American Midland Naturalist* 18 (5): 830–841.

*Wilson, J., L. Wilson, and J. Nicholas. 1987. *Wildflowers of Yosemite*. Sunrise Productions, Yosemite, Calif.

*Winterringer, G. S. 1967. *Wild Orchids of Illinois*. Illinois State Museum, Springfield.

Wood, J. 1986. *Calypso bulbosa* var. *occidentalis* and var. *speciosa*. *Kew Magazine* 3: 147–151.

Index

Page numbers in **boldface** indicate the main text descriptions. The color plates follow page 108.